AF822522

openstax™

High School Physics Lab Manual
Student Version

XanEdu

OpenStax
Rice University
6100 Main Street MS-375
Houston, Texas 77005

To learn more about OpenStax, visit https://openstax.org.
Individual print copies and bulk orders can be purchased through our website.

©2020 Rice University. Textbook content produced by OpenStax is licensed under a Creative Commons Attribution 4.0 International License (CC BY 4.0). Under this license, any user of this textbook or the textbook contents herein must provide proper attribution as follows:

- If you redistribute this textbook in a digital format (including but not limited to PDF and HTML), then you must retain on every page the following attribution: "Access for free at openstax.org."
- If you redistribute this textbook in a print format, then you must include on every physical page the following attribution: "Access for free at openstax.org."
- If you redistribute part of this textbook, then you must retain in every digital format page view (including but not limited to PDF and HTML) and on every physical printed page the following attribution: "Access for free at openstax.org."
- If you use this textbook as a bibliographic reference, please include https://openstax.org/details/books/physics in your citation.

For questions regarding this licensing, please contact support@openstax.org.

Trademarks
The OpenStax name, OpenStax logo, OpenStax book covers, OpenStax CNX name, OpenStax CNX logo, OpenStax Tutor name, Openstax Tutor logo, Connexions name, Connexions logo, Rice University name, and Rice University logo are not subject to the license and may not be reproduced without the prior and express written consent of Rice University. XANEDU and the XanEdu Logo are trademarks of Xanedu Publishing, Inc. The XANEDU mark is registered in the United States, Canada, and the European Union. These trademarks may not be used without the prior and express written consent of XanEdu Publishing, Inc.

PAPERBACK BOOK ISBN-13 **9781711493312**
ORIGINAL PUBLICATION YEAR **2020**
10 9 8 7 6 5 4 3 2 1

Printed by
XanEdu
4750 Venture Drive, Suite 400
Ann Arbor, MI 48108
800-562-2147
www.xanedu.com

OpenStax

OpenStax provides free, peer-reviewed, openly licensed textbooks for introductory college and Advanced Placement® courses and low-cost, personalized courseware that helps students learn. A nonprofit ed tech initiative based at Rice University, we're committed to helping students access the tools they need to complete their courses and meet their educational goals.

Rice University

OpenStax, OpenStax CNX, and OpenStax Tutor are initiatives of Rice University. As a leading research university with a distinctive commitment to undergraduate education, Rice University aspires to path-breaking research, unsurpassed teaching, and contributions to the betterment of our world. It seeks to fulfill this mission by cultivating a diverse community of learning and discovery that produces leaders across the spectrum of human endeavor.

RICE

Philanthropic Support

OpenStax is grateful for our generous philanthropic partners, who support our vision to improve educational opportunities for all learners.

Laura and John Arnold Foundation
Arthur and Carlyse Ciocca Charitable Foundation
Ann and John Doerr
Bill & Melinda Gates Foundation
Girard Foundation
Google Inc.
The William and Flora Hewlett Foundation
Rusty and John Jaggers
The Calvin K. Kazanjian Economics Foundation
Charles Koch Foundation
Leon Lowenstein Foundation, Inc.
The Maxfield Foundation
Burt and Deedee McMurtry
Michelson 20MM Foundation
National Science Foundation
The Open Society Foundations
Jumee Yhu and David E. Park III
Brian D. Patterson USA-International Foundation
The Bill and Stephanie Sick Fund
Robin and Sandy Stuart Foundation
The Stuart Family Foundation
Tammy and Guillermo Treviño

To the Student:

The experiments in this lab manual were designed to enhance your learning of physics through hands-on learning. Here, you will use actual physics lab equipment and scientific research methods that will help you connect what you learn in physics with your everyday life! You do not need any previous experience with physics to successfully perform these lab procedures. However, reading this preface will help you get the most out of your laboratory experience. **If you are ever unsure about a procedure, please ask your teacher before proceeding.** Improperly performing lab procedures can not only result in poor data and having to redo experiments, but could damage equipment or injure you or your classmates.

Components of a Lab

Main introduction:
Each lab contains an introductory section under the title. This introduces the "big picture" concepts of the lab as well as how they connect to everyday life. They will also introduce the pioneering physicists and experiments that led to our current knowledge of each lab topic. Relevant equations that you will use in the labs are also introduced here, including definition of their variables. Many of the labs involve measuring the value of these variables so that you can later perform your own calculations. Please read the lab and activity introductions carefully before your lab period. Then, ask your teacher about any concepts that you are unsure of before the lab starts.

In this lab you will learn:
This section presents learning objectives for the lab. These are the "take away points" that you should be able to explain or perform after doing the activities. It is helpful to read these objectives before each lab to prime yourself for what you will learn. It is then helpful to reread these at the end of each lab to ensure that you have achieved all of the learning objectives.

Activities:
Each lab is divided into 2-3 activities. Please note that your teacher may or may not have you perform all activities in a given lab so pay close attention to your teacher's instructions throughout the lab.

Safety precautions:
These bullet points list important safety issues that will prevent the injury to yourself or your classmates during the lab activities. Each activity has its own safety precautions section. **Please read and understand all safety precautions before beginning each activity!**

"For this activity you will need…" section:
This section lists all of the materials needed for each activity. Before you start the lab, make sure that you can identify all items on this list. Also, pay close attention to your teacher's instructions, as you may be using different equipment for these labs then those on this list.

Activity introduction:
These are short introductions relevant to specific activities. As with the main introduction, the activity introductions may contain formulas, equations or other background information that needed to successfully carry out and understand the activities. As with the main introduction, please read these introductions carefully before your lab period, and ask your teacher about any material that you are unsure of.

Process steps:
These are the steps you will perform to carry out the activities. Please read through **all** of the process steps and setup diagrams **before** starting Step 1. Ask your teacher if there are any steps you don't understand prior to starting. This will help you perform the activities correctly the first time, preventing the need to redo activities or having to leave your laboratory period with unusable data.

Data and observations:
This section provides all of the tables and other space you will need to record your data. Note the units used in the tables and ensure that you measure in the same units. Write neatly and legibly so that you will understand your data at a later time. If you need more space, copy the tables into your notebook. Follow your teacher's instructions closely regarding recording and turning in data.

Assessments:
The assessment sections provide questions that test your knowledge of the lab material. Your teacher will instruct you on how to submit answers to the assessments for grading.

LO and TEKS tags:
Texas Essential Knowledge and Skills (TEKS) tags are applied to various parts of the lab. The TEKS are defined on the Texas Education Agency website (your teacher will also have a list of the TEA standards).

Both types of tags are provided in brackets within the activity titles. You can use these tags to quickly find and review lab materials relevant to the TEA standards. In addition, TEKS tags are found with the assessments, allowing you to review assessments relevant to specific TEA standards.

Table of Contents

Lab 1: Measurement, Precision, and Accuracy

Physics, like other fields of science, is based on observation and experiment. A wide diversity of measurements is routinely used in physics, including measurements of length, volume, mass, and temperature. Making these measurements depends heavily on the ability to make accurate and precise measurements.

Suppose you use a ruler to measure the length of a piece of string marked as 30.48 cm (or 0.3048 m). **Accuracy** refers to how close a measurement is to the correct value for that measurement. If you measure the string three different times and all three measurements are very close to 30.48 cm, then the measurements are accurate. **Precision** refers to the difference between a group of repeated measurements, also known as the *spread* of the measurements. Precision also considers the range between measurements as well as how often you get repeated measurements. For example, if the measurements of the string do not vary drastically, then they are precise.

<u>In this lab you will learn:</u>

- how to measure volume using the displacement method;
- how to measure mass using a triple beam balance, spring scale, and electric balance;
- how to measure distance using rulers, meter sticks, and string.

Activity 1: Measuring Volume (TEKS 2H; 2J)

Volume is defined as the amount of space that a substance or object occupies, or the amount of space enclosed within a container. In this activity, you will be measuring the volumes of three masses (of known volume) using a graduated cylinder. You will also learn how to measure volume using the **displacement method**, in which an object is placed into a measured container with a known volume of water. The fluid level changes when the object is fully immersed, indicating the volume of the object. The displacement method is helpful for measuring volume when an object is oddly shaped and it is hard to determine the dimensions of an object.

Significant figures include figures read from the measuring instrument plus one **doubtful digit,** which is retained and estimated by the observer. To determine the number of significant figures for a measuring instrument, evaluate the resolution of the instrument to get the known digits. Nonzero digits are always considered significant. However, zeros can function as placeholders and are considered significant if they are surrounded by nonzero digits on both sides. Zeroes to the right of significant figures, as in the case of 1000, are not significant unless they are used to the right of a decimal, as in the case of 2.00. Zeroes to the left of significant figures, as in the case of a decimal such as 0.005, are not considered significant unless there is a nonzero digit to the left of the decimal, as in the case of 3.004.

Safety Precautions

- Place objects gently into the graduated cylinder to avoid breakage.
- Inform your teacher immediately of any broken glassware, as it could cause injuries.
- Clean up any spilled water or other fluids to prevent other people from slipping.

For this activity you will need the following:

- Water
- Solid objects that will sink
- Ruler marked in centimeters (cm)
- Graduated cylinder

For this activity, you will work *in pairs*.

Procedure

Step 1: Measure the volume of the three objects given to you by your instructor, using the height, length, and width of the object and the formula for volume (known as the formula method). Record your results in Table 1.1 under *Measured Volume.*

Step 2: Use the displacement method to measure the volume of the three objects (Figure 1.1). Record your results in Table 1.1 under *Displacement Volume.*

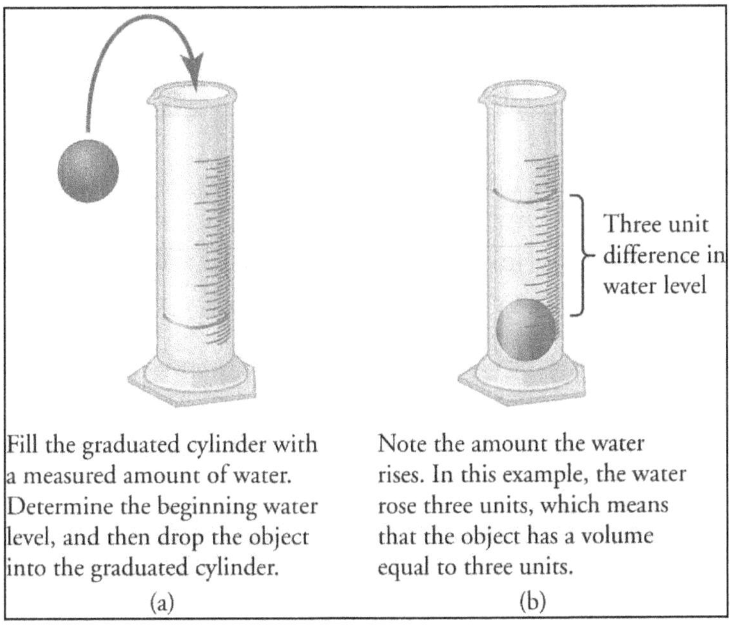

***Figure 1.1:** The displacement method can be broken down into two steps: (a) Place the object into the cylinder with a premeasured amount of water and (b) measure the number of volume units the water rises. Note that the displacement in this figure is not accurate and is for illustrative purposes only.*

Consider the meniscus when reading the volume of the water. The **meniscus** is the curve in the surface of the water that results when water interacts with glass (Figure 1.2).

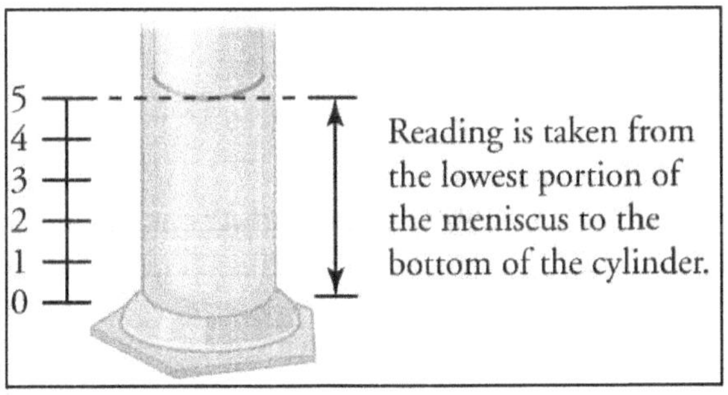

***Figure 1.2:** It is important to take the meniscus into account when measuring the volume of a liquid in a graduated cylinder. Remember to record the lowest point of the meniscus.*

Step 3: Determine the accuracy and precision of the measurements you recorded in Steps 1 and 2 and record them in Table 1.2.

- Accuracy can be measured using percent error. Here, the accepted value is a known value, e.g., the exact volume of a ball or length of a string. The experimental value is the value that was actually measured.

$$\% \text{ error} = \frac{\text{accepted} - \text{experimental}}{\text{accepted}} \times 100\%$$

- Precision can simply be measured as the difference between an obtained value and the average value

Precision = average – obtained.

The more accurate method will be closest to the actual measurement; the more precise method will have the least spread.

Data and Observations

Number of significant figures available in graduated cylinder:

Table 1.1: Volume

Object	Known Volume	Displacement Volume	Measured Volume
Water			
Object 1			
Object 2			
Object 3			

Table 1.2: Precision and Accuracy

	Accuracy	Precision
Formula		
Displacement		

Assessments

1. A student measures an object that is known to be 3.0 cm^3 in volume multiple times and records values of 3.1 cm^3, 2.9 cm^3, and 2.8 cm^3. [TEKS 2H]
    a. What would allow these measurements to be considered accurate? Explain your answer.
    b. What would allow these measurements to be considered precise? Explain your answer.
2. Accuracy refers to how close a measurement is to the ... value for that measurement. [TEKS 2H]
3. Based on your results, is measuring volume by formula or displacement more accurate? Which is more precise? Cite evidence from your results from this lab. [TEKS 3B]
4. An object of unknown volume is placed into a graduated cylinder filled with a measured amount of water. The object causes the water to rise by 5 units. [TEKS 2J]
    a. What method is being used to measure the volume?
    b. What is the volume of the object?

Activity 2: Measuring Mass

Mass is a measure of the amount of matter contained in an object. Mass is not the same as weight; weight is the measurement of gravity's pull on an object. Mass is not affected by the planetary location of an object—your mass is the same if you are standing on Earth or if you are standing on Earth's moon. However, you weigh about one-sixth as much on the Moon as you do on Earth.

Mass is measured using a balance that compares a known amount of matter to an unknown amount of matter. You will use a **triple beam balance**, **spring scale**, and **digital scale** to determine the mass. The spring scale measures the force the object is exerting on the scale's hook. You will also calculate the volume of each object to determine its density. **Density** is defined as mass per unit volume.

A spring scale measures the force exerted on an object, in newtons (N). You will use it to measure the force of gravity on the weights. Hang an object on the hook of the scale and read the scale at the top of the platform to determine the number of newtons exerted by gravity on the object.

Safety Precautions

- Be sure you are carrying the triple beam balance properly; place one hand under the stage base and the other hand under the balance pointer column.
- Do not try to carry too many materials at once; make multiple trips if needed.
- Do not force the adjustment knob on the triple beam balance.

For this activity you will need the following:

- Triple beam balance
- Spring scale
- Digital scale
- Solid objects: standard cube or rectangular weights that can be measured for density calculations
- Ruler marked in centimeters (cm)

For this activity, you will work *in pairs*.

Procedure

Step 1: Put the triple beam balance on a flat, horizontal surface and push all the riders toward the metal platform so they each are on zero (Figure 1.3).

Step 2: Check the pointer on the right side; it should be on zero. If the pointer is still moving, wait for it to stop, and make sure it's exactly at zero.

Step 3: If the pointer is above or below zero, turn the adjustment knob slightly so that the pointer hits zero. Clockwise turns make the pointer move down and counterclockwise turns make the pointer move up.

Step 4: Place the first object on the scale. Move the appropriate slider until the pointer approaches zero. Start with the biggest slider and decrease to the smallest. If the biggest slider causes the pointer to go below zero, move the larger slider back and select the next smaller slider. Continue with the sliders until the pointer swings the same small amount above and below the zero marker.

Step 5: Read and record the mass from the scale in Table 1.3. Continue with the other two masses.

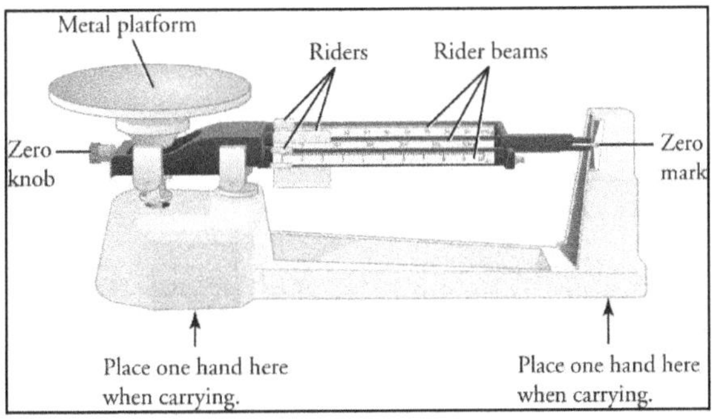

Figure 1.3: *A triple beam balance measures the mass of an object by comparing it with the known masses of riders attached to three horizontal beams.*

Step 6: Hold up the spring scale (Figure 1.4). Check to see that the midline of the pointer is at zero. If it is not at zero, adjust the knob or screw at the top of the spring scale so that the platform is at zero. Confirm with your teacher that your spring scale is zeroed properly.

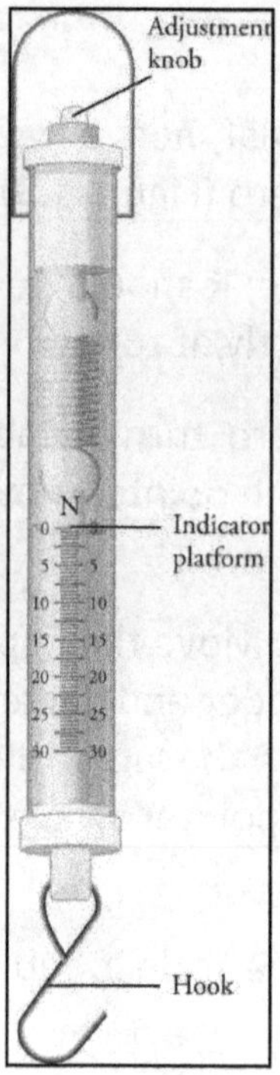

***Figure 1.4:** A spring scale measures the weight of an object by comparing it to the known weight that causes the same tension in the spring.*

Step 7: Read the newtons exerted by gravity on the object and record the number in Table 1.3. Repeat with the remaining objects.

Step 8: Pull out the digital scale and hit *tare* so the scale reads zero. Place each of the objects on the provided digital scale. Record the masses in Table 1.3.

Data and Observations

Number of significant figures available in triple beam balance: ....

Number of significant figures available in spring scale: ....

Number of significant figures available in digital scale: ....

Table 1.3: Mass Measurements

	Triple Beam Balance Mass (g)	Spring Scale Weight (N)	Digital Scale Mass (g)
Object 1			
Object 2			
Object 3			

Assessments

1. The kilogram (kg) is the base unit for mass in the International System of Units (SI). The measurements for the triple beam balance and digital scale are in grams. Convert your results in Table 1.3 to kg using scientific notation. [TEKS 2H]
2. To compare and contrast the differences in precision and accuracy for each measurement, the numbers need to be in the same units. How would you convert the number of newtons determined from the spring scale into kg? Convert the force exerted on the spring scale into kg. [TEKS 2H]
3. Which method is more accurate? Which is more precise? [TEKS 2H]

Activity 3: Measuring Distance

Many calculations in physics depend on distance measurements that describe the positions of objects in space. **Distance** is the length of a line between two points. We use distance in physics to determine velocity, acceleration, and other important values. It is important to develop methods for measuring distances that are precise, accurate, and in units we can use. Consider the **significant figures** for a meter stick. What will be the **doubtful digit** when length is measured with a meter stick from your classroom?

Safety Precautions:

- Be mindful of your surroundings when using a meter stick to avoid collisions.

For this activity you will need the following:

- Small block
- Ruler marked in cm
- Meter stick

For this activity, you will work *in pairs*.

Procedure

Step 1: Look closely at the ruler and meter stick provided. Note that it has two sides; you will use the metric side for this lab. The entire length of the meter stick is 1 meter; this is the SI standard unit for distance. The meter stick is broken down into increments of 1 centimeter, and each centimeter is broken down into millimeters. Given the resolution of the meter stick, what is the number of significant figures for the meter stick? To calculate the significant figure, determine the number of digits shown on the instrument, and then add one figure as an approximation. Record your answer below:

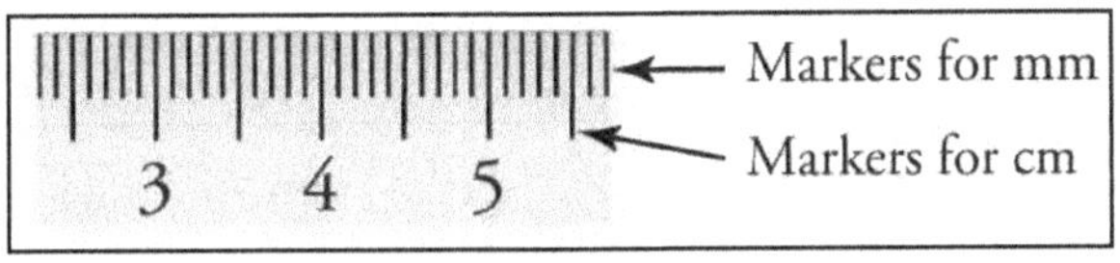

Figure 1.5: *You can see that the metric side of a ruler measures millimeters and centimeters.*

Step 2: Use the ruler to measure the length, width, and height of the provided small block in the number of significant figures determined in Step 1; record in Table 1.4.

Step 3: Use the meter stick to measure the length, width, and height of your lab desk in the number of significant figures determined in Step 1; record in Table 1.4.

Step 4: Use the meter stick to record the length and width of your classroom in the number of significant figures determined in Step 1; record in Table 1.4. Ask your teacher for the height of your classroom.

Data and Observations

Number of significant figures available in a standard meter stick: ....

Table 1.4: Measuring Distance

	Length	Width	Height
Small block			
Desk			
Classroom			

Assessments

1. Were any of your measurements obtained in cm or mm? Develop and use a table to convert the obtained measurements into the SI base unit for distance (m). Use scientific notation if applicable. [TEKS 2H/2J]
2. Do any of the measurements obtained show any uncertainties? Is there a way to ensure greater accuracy of any of the uncertain measurements? [TEKS 2I]
3. If you can find a way to measure the height of the room, do so. Does it confirm the height provided by your teacher? If there is no way to measure the height directly, is there another way to find the height? [TEKS 2I]

Lab 2: Position and Speed of an Object

Quantifying the motion of objects is a key part of physics. Many objects move in complicated ways, but the tools we use to describe *complicated* motion are the same as for *simple* motion. Before you continue your study of motion, it is important to define and understand these analysis tools.

A moving object starts its motion at one point and ends its motion at another point. The **distance** an object travels—measured in meters in the SI system—is how far the object has gone between the start and end points of its motion, along the exact path taken. Distance is different from **displacement**; displacement is where the object is relative to where it started, but does not describe the *straight path* that the object took to get from the start to the end. The object's **speed** describes how quickly the object gets from one point to another. An object that is not speeding up or slowing down is moving at a constant speed, meaning that the speed calculated at any point along its path is the same. The average speed of an object can be calculated using

$$\text{average speed} = \frac{\text{distance traveled}}{\text{time traveled}} = \frac{d}{t}.$$

<u>In this lab you will learn:</u>

- how to calculate the speed of an object using measured distance and time values;
- how to draw and understand a line of best fit for a plot line;
- how to determine the speed of an object using a distance vs. time graph and best fit line.

Activity 1: Calculating Speed Using Distance and Time TEKS (4)(B)

In this part of the lab, you will determine the speed of an object by using the formula definition of speed. Distance is measured using a meter stick and is reported in meters. Time is measured using a stopwatch and is reported in seconds in the SI system. The units for the calculated speed are therefore meters per second.

Safety Precautions

- Do not push the wind-up car too fast. Keep its motion under control.
- Keep the car moving in a straight line and aim it in the right direction to avoid crashing.
- Avoid horseplay, such as racing cars over table edges or swinging meter sticks.

For this activity you will need the following:

- Wind-up car
- Meter stick
- Tape
- Stopwatch

For this activity you will work *in pairs.*

Procedure

Step 1: Choose a starting place on the floor about 20 cm from the wall for your car with plenty of space ahead. Mark the starting point with a piece of tape approximately 10 cm long. Use the middle of this piece of tape as the starting line for your car.
Step 2: Starting from the point you marked in Step 1, measure half of a meter (50 cm) and place another piece of tape parallel to the first. Repeat this three more times so you have pieces of tape at 50 cm, 100 cm, 150 cm, and 200 cm from the starting line. See Figure 2.1. Make all measurements from the middle of the pieces of tape.
Step 3: Wind up the car and hold it in place at the starting line. Keep track of how much you wound the car so you can wind it the same amount in all trials.
Step 4: Countdown "3, 2, 1, go!" On *go*, one partner should release the car.

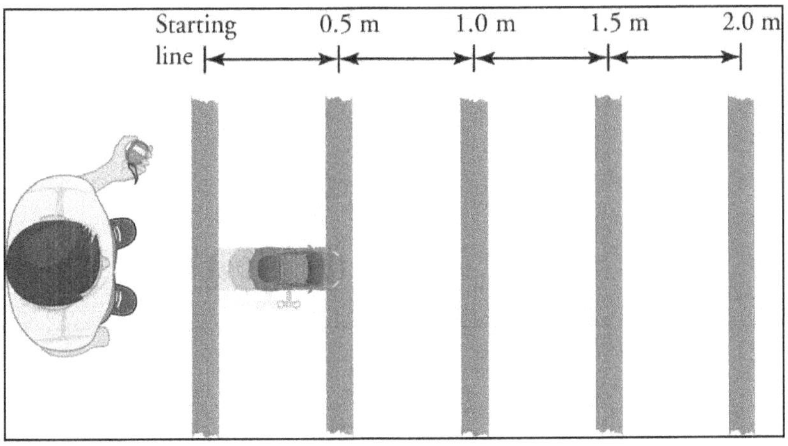

Figure 2.1: *The speed of a car can be calculated by timing how long it takes the car to travel a fixed, pre-measured distance. At a constant speed, it should take the same amount of time for the car to travel between each marker.*

Step 5: The other partner should *start* the stopwatch when the front of the car crosses the first piece of tape. When the front of the car crosses the last piece of tape at 2.0 meters, *stop* the stopwatch. Record in Table 2.1 the amount of time it took for the car to travel 2.0 meters.
Step 6: Repeat Steps 3–5 two more times and record the measured times in Table 2.1.
Step 7: Using the measured distances and times, calculate the speed of the car for each trial and record the value in the last column of Table 2.1.

Data and Observations

Table 2.1: Calculated Speed from Distance and Time

Trial	Distance (m)	Time (s)	Speed (m/s)
1	2.0		
2	2.0		
3	2.0		

Are the speeds you calculated in the third column of your data table the same? Should they be the same, and why? Explain why there might be some variation in the speed.

Assessments

1. A student using a stopwatch finds that it takes his dog 15 s to run across his yard, which is 30 m wide. TEKS (4)(B)
    a. What is the dog's average speed?
    b. Assuming that the dog is running at the same speed, how long would it take the dog to run across an 80-m wide yard?
2. A student times a car traveling a distance of 2 m. She finds that it takes the car 5 s to travel the first meter and then another 8 s to travel the second meter. Is the car traveling at a constant speed? How do you know? TEKS (4)(B)
3. Speed is defined as distance divided by time. In this lab, the unit of distance was meters and the unit of time was seconds. Therefore, the unit of speed was meters per second. Is miles per hour also a unit of speed? What about centimeters per minute? Why or why not?

Activity 2: Graphing to Determine Speed TEKS (4)(A)

In this part of the lab you will determine the speed of an object by graphing. The slope of a graph is calculated by

$$m = \frac{\Delta y}{\Delta x}.$$

If the *y*-axis of the graph is distance and the *x*-axis of the graph is time, then Δy is the distance traveled and Δx is the amount of time it took. Then, the slope of the distance versus time graph is equal to the object's speed

$$m = \frac{d}{t} = v.$$

For an object traveling at a constant speed, you would expect that if you measured the time taken to go any distance, the calculated speed should be the same. However, every measurement has some error, so you will get a slightly different speed each time. Using a **line of best fit** helps to average over these errors and give a more accurate estimate of the calculated speed. To draw a line of best fit, use a ruler to draw a line that is as close to all of the data points as possible and follows the trend of the data. Some points will lie above the line and others below the line due to measurement errors. However, because the best fit line goes through the middle of the data points, it averages the values. As a result, the slope of the line of best fit is a more accurate measurement of speed than using any individual data point.

Safety Precautions

- Do not push the wind-up car too fast. Keep its motion under control.
- Keep the car moving in a straight line and aim it in the right direction to avoid crashing.
- Avoid horseplay, such as racing cars over table edges or swinging meter sticks.

For this activity you will need the following:

- Wind-up car
- Meter stick
- Tape
- Stopwatch

For this activity you will work *in pairs*.

Procedure

Step 1: Wind-up the car and hold it in place behind the starting line, as set up in the first activity of this lab.
Step 2: Countdown "3, 2, 1, go!" On *go*, one partner should release the car.
Step 3: The other partner should *start* the stopwatch when the front of the car crosses the first piece of tape. When the car crosses the first piece of tape at 0.5 meters, *stop* the stopwatch. Record in Table 2.2 the amount of time it took for the car to travel 0.5 meters.
Step 4: Repeat Steps 2 and 3 for when the car crosses the next three pieces of tape, corresponding to distances of 1.0 m, 1.5 m, and 2.0 m.
Step 5: On the graph in Figure 2.2, add appropriate scales to the *x*-axis *Time (s)* and the *y*-axis *Distance (m)* based on the data in Table 2.2.
Step 6: Plot the distance versus time data from the table on the axes provided.
Step 7: Draw a line of best fit for the data on your plot. Calculate the slope of the line of best fit and record the value, with appropriate units, on the space provided.

Data and Observations

Table 2.2: Time of Travel for Given Distances

Trial	Distance (m)	Time (s)
1	0.5	
2	1.0	
3	1.5	
4	2.0	

For each provided distance, record the time taken for the car to travel that distance.

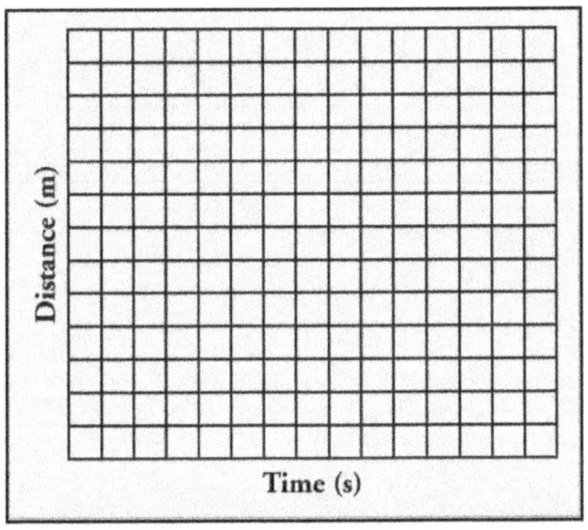

Figure 2.2: *A distance versus time plot for an object moving at a constant speed produces a graph whose slope is the object's speed. By plotting data points and drawing a line of best fit to the data, a value for the object's average speed can be obtained.*

Assessments

1. Is it correct to use the coordinates of any data point on your plot to calculate the slope of a line of best fit? Why or why not? TEKS (4)(A)
2. In the activity, you plotted your data with distance on the *y*-axis and time on the *x*-axis. If you switched the axes when plotting the data, what would be the units of the slope? Would the slope still represent the speed? TEKS (4)(A)
3. In the graph of your data from the activity, would it make sense for your line of best fit to pass through (0, 0)? Why or why not? TEKS (4)(A)

Lab 3: Acceleration

Objects in motion have many features we can describe based on the distance they travel and the time it takes to get there. One quantity you may have already encountered is *velocity*, which is determined by dividing an object's displacement by the total time of travel. However, as you know from experience, many objects do not travel at the same velocity all the time—objects speed up and slow down, and change direction.

In physics, we describe speeding up and slowing down by looking at the change in velocity and dividing it by the amount of time it takes to make the change. This quantity is referred to as an object's **acceleration**. These characteristics of an object's motion can be represented using graphs, which provide a good tool for comparing the motions of different objects. In this lab, you will collect and analyze data of objects in motion both arithmetically and visually.

In this lab you will learn:

- how to gather and use distance and time data to determine an object's acceleration through space;
- how to create a graph to represent motion in terms of an object's velocity over time and use it to find an object's acceleration.

Activity 1: Calculate Acceleration TEKS (4)(B

Acceleration is a change in velocity over a period of time. If you know the velocity of an object at two points as well as the time it took to travel between them, you can then calculate the acceleration. Recall that velocity is calculated using the equation

$$\text{velocity} = \frac{\text{final distance} - \text{initial distance}}{\text{final time} - \text{initial time}}.$$

Similarly, acceleration is calculated as

$$\text{acceleration} = \frac{\text{final velocity} - \text{initial velocity}}{\text{final time} - \text{initial time}}.$$

Photogates are able to detect when a light beam shining between the two ends of the photogate is blocked. The amount of time the light is blocked is a data point. See Figure 3.1. If the light of two photogates is blocked and unblocked in succession, the interval of time it takes the cart to travel between the two photogates is recorded.

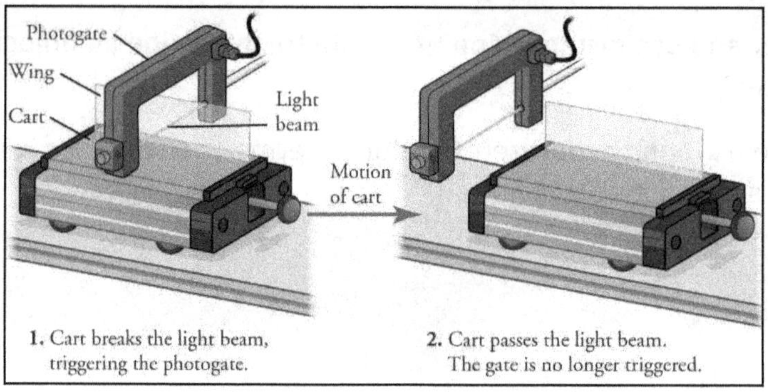

Figure 3.1: *The photogate is triggered when the light beam between the two ends of the gate is broken (Step 1) and is no longer triggered when the beam is reconnected (Step 2).*

Safety Precautions

- Be gentle with the electronics; sensors on the photogates must be handled carefully.
- Use the carts as directed—make sure that the carts do not fall on the ground, collide with each other, or strike anyone.

For this activity you will need the following:

- One cart
- Two photogates
- Ramp
- Ruler in cm
- Device to capture photogate data, such as a graphing calculator
- Protractor

For this activity you will work *in pairs*.

Procedure

Step 1: Your teacher will provide instruction on how to set up and use your specific brand of photogates. Pay close attention. Once you fully understand the operation of your photogates, set up your ramp with the two gates some distance apart. Be sure to set the photogate at the top of the ramp far enough down the track so that it is not triggered by the cart at rest.

Step 2: Measure the angle between the ramp and the table. Record in Table 3.1.

Step 3: Measure and record in Table 3.2 the distance from the stop at the top of the ramp to the first photogate.

Step 4: Make sure the photogates are set to run in Gate mode.

Step 5: Place your cart at the top of the ramp so that it lightly touches the stop at the top of the ramp. Release the cart, without applying additional force, so that the cart rolls down the ramp. As shown in Figure 3.2, during Gate mode the wing of the cart triggers the photogate to start timing when it first blocks the light. The gate is triggered to stop timing when the light beam is no longer broken.

Step 6: Record the velocity of the cart at photogate A and photogate B in Table 3.2.

Step 7: Repeat Steps 5 and 6, recording the data in Table 3.2. You should end up having two values for the velocity at photogate A and two values for velocity at photogate B.

Step 8: Set the photogates to run in Pulse mode. Keep the photogates set up as they were before. Now that you have the velocities at both photogates, you need to find the time it takes to travel between the photogates. This equals the time it takes to go from the velocity traveled at photogate A to the velocity traveled at photogate B.

Step 9: Again, place the cart at the top of the ramp so it lightly touches the stop. Release the cart without applying additional force, so the cart rolls down the ramp. Try to release the cart down the ramp the same way each time.

Step 10: Record the time it takes for the cart to pass through both photogates in Table 3.2.

Step 11: Repeat Steps 8 through 10. You should end up having two values for the time it takes to travel between photogate A and photogate B.

Step 12: Move the photogates. Repeat Step 3 through Step 11 until you have completed Table 3.2.

Step 13: Complete Table 3.3. The Trial 1 Averages are calculated using the Trial 1.1 and 1.2 data to fill in the *Trial 1 Average* row, and so on. Calculate the acceleration for each gate configuration using the averages.

Step 14: Compare the experimental accelerations. Are they close to each other? Why or why not?

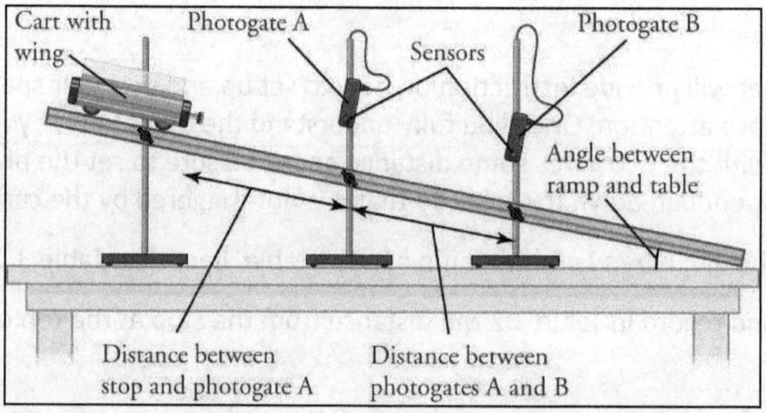

Figure 3.2: *The setup depicted here will allow you to experiment with acceleration.*

Data and Observations

Table 3.1: Setting Up the Ramp

Angle between the ramp and table	

Table 3.2: Data Collected from the Photogates

	Distance from the top of the photogate (cm)	Velocity at photogate A (cm/s)	Velocity at photogate B (cm/s)	Time interval between photogates A and B (s)
Trial 1.1				
Trial 1.2				
Trial 2.1				
Trial 2.2				
Trial 3.1				
Trial 3.2				

Table 3.3: Data Averages and Mathematical Analysis of Data

	Distance from the top of the photogate (cm)	Velocity at photogate A (cm/s)	Velocity at photogate B (cm/s)	Time interval between photogates A and B (s)	Acceleration between photogates A and B (cm/s^2)
Trial 1 Average					
Trial 2 Average					
Trial 3 Average					

Assessments

1. Describe any differences you notice in your data regarding velocity and acceleration as the distance from the top of the ramp changed. TEKS (4)(B) TEKS (2)(J)
2. The accepted value for the acceleration at any given point on a ramp, assuming no additional forces are acting upon the cart, is

   $$\text{acceleration} = g \times \sin(\text{angle of ramp relative to table})$$

   Note—This assumes the table is level.

   Measure the angle of your ramp and calculate the theoretical acceleration for your set up.

   Compare the average of the acceleration values to this number by calculating the percent error using the equation

   $$\text{Percent error} = \frac{\text{accepted acceleration} - \text{experimental value}}{\text{accepted acceleration}} \times 100 \text{ percent}$$

   What are some potential sources of error? TEKS (4)(B)

3. At what point might the average velocity match the instantaneous velocity of the cart? If the acceleration of the cart changed as it traveled down the ramp, would this have an effect on when the average velocity matches the instantaneous velocity? TEKS (4)(B)

Activity 2: Graphing Acceleration TEKS (4)(A), (2)(J)

Graphing velocity versus time can give a summary of how an object moves through space. See Figure 3.3. When graphing velocity versus time, the *y*-intercept gives the velocity of an object when time is zero. Often, we set time equal to zero for our first data point and plot the data from there. When discussing the slope of a velocity versus time graph, you are looking at the change in velocity (*rise*) over change in time (*run*). Because the difference in velocity over the difference in time is acceleration, the slope of a velocity versus time graph gives the acceleration of the object. A negative value for acceleration would indicate a decrease in velocity while a positive value for acceleration would indicate an increase in velocity.

For this activity you will need the following:

- Pencil
- Ruler

For this activity you will work *alone*.

Procedure

Note—For this procedure, you will use the data in Table 3.4.

Step 1: Label the horizontal axis of your graph *Time (s)*; note the inclusion of the appropriate units as determined by the data.

Step 2: Label the vertical axis of your graph *Velocity (m^2/s)*; note the inclusion of the appropriate units as determined by the data.

Step 3: Determine the scales of your axes by first looking at the number of grid lines on each axis and comparing them to the span of numbers you need to accommodate on each axis. Label the gridlines on each axis appropriately.

Step 4: Graph the data points on Figure 3.4. Remember, move along the horizontal axis to represent the time value and then up to represent the velocity value.

Step 5: Draw a best-fit line to follow the trend of your data. This is done by making sure you go through as many points as possible while having about the same number of points on each side of the line.

Step 6: Give your graph an appropriate, descriptive title that tells the reader what the graph represents.

Step 7: Determine the *y*-intercept and slope of the line, remembering to include units, and interpret what they mean with respect to the object in motion.

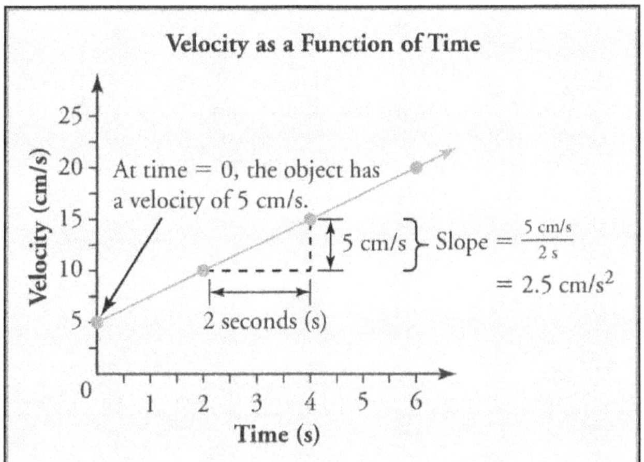

***Figure 3.3:** We know that the object has an initial velocity of 5 cm/s because at the initial time, $t = 0$, that is the value of velocity. We know acceleration is given by looking at rise over run, or the change in velocity over the change in time, the slope of the velocity versus time graph. In this case, we have 5 cm/s (rise) divided by 2 s (run), resulting in an acceleration of 2.5 cm/s^2.*

Data and Observations

The following data were collected when clocking a cheetah as it was hunting.

Table 3.4: Data Collected of Cheetah Hunting in the Wild

Velocity (m/s)	Time (s)
0	1
6.5	1.5
13	2
19.5	2.5
26	3

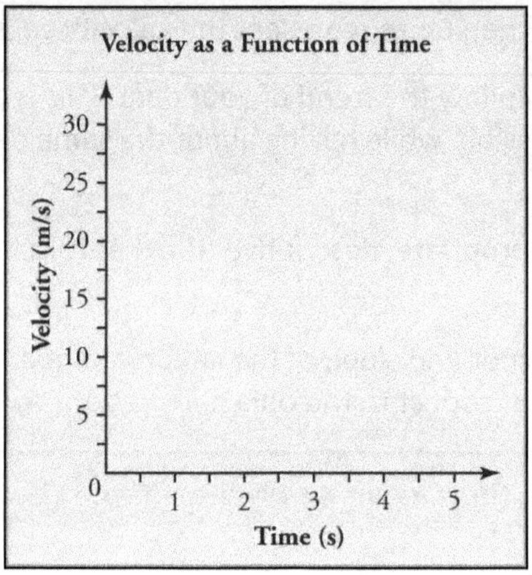

Figure 3.4: *Axes for velocity versus time graph.*

Assessments

1. If the graph of a bike's velocity versus time has a negative slope, what is the bike doing in terms of acceleration? Is acceleration positive or negative? Is the velocity increasing or decreasing? How do you know? TEKS (4)(A) TEKS (2)(J)
2. If the plot line of an object's velocity versus time remains parallel to the *x*-axis for three seconds, how would you describe the object's motion during those three seconds? What is the object's acceleration? What does that tell you about the object's velocity over those three seconds? What does this tell you about the object's position over the time period? TEKS (4)(A) TEKS (2)(J)
3. Suppose you have a car that can accelerate from 0 to 31 meters per second in four seconds. Describe how the slope of a velocity versus time graph of the car's motion compares to the graph of the cheetah's velocity versus time data. TEKS (4)(A) TEKS (2)(J) TEKS (2)(L)

Lab 4: Newton's Laws

Newton's first law tells us that an object at rest tends to stay at rest, and an object in motion tends to stay in motion, unless acted upon by an unbalanced force. What is a force? A force is a push or a pull on an object that results from an interaction with another object. You experience forces every day—running to catch a bus, riding a bike, and driving a car. In all of these cases, forces are acting upon you. When a force acts upon an object, the object accelerates. Force is measured in newtons, or

$$\mathrm{kg \cdot m/s^2}\ .$$

In this lab, you will examine the force on objects resulting from gravity. This force is what we customarily call weight. The force due to gravity, or weight, can be calculated directly from the mass *m* and from the known constant acceleration due to gravity at the surface of Earth

$$g = 9.8\ \mathrm{m/s^2}\ .$$

Newton's second law goes on to say that the acceleration of an object is directly proportional to the net force applied to it and inversely proportional to the mass of the object. In other words, the harder you push, the more an object will accelerate; and, the heavier an object is, the less it will accelerate. This relationship is expressed in the equation

$$F_{net} = ma\ .$$

We can calculate the force of gravity, or weight, with the following formula

$$F_{\mathrm{g}} = mg\ .$$

As you have learned, acceleration is defined as the change of velocity per unit time, and is the slope of a velocity vs. time graph.

Also in this lab, you will explore how net forces create the push or pull on an object or system, causing acceleration. You will calculate the acceleration you might expect to see, knowing the masses of the objects in the system and the force of gravity, and compare that calculation to the results you measure in the experiment.

<u>In this lab you will learn:</u>

- how to use free-body diagrams to determine and visualize experimental variables for force and motion;
- how to calculate the force exerted on a mass due to gravity;
- how to calculate acceleration from force and mass;
- how to calculate acceleration from measured changes in velocity over a period of time.

Activity 1: Drawing a Free-Body Diagram and Calculating Force of Gravity
TEKS (4)(B), (4)(D), (4)(E)

Drawing a picture of the forces acting on an object can help you visualize what is happening in a system and allow you to make better hypotheses. We call this drawing a **free-body diagram**. You will use the image of the experimental setup shown in Figure 4.1 to draw a free-body diagram. This experimental setup will be used in Parts 2 and 3 of this lab.

Safety Precautions

None

For this activity you will need the following:

- Pencil
- Calculator
- Lab setup image, see below

For this activity you will work *alone*.

Procedure

Note—For this procedure, you will be using the diagram in Figure 4.1. Ignore air resistance.

Step 1: The diagram shows two masses: m_1 is the total mass of the cart and any added load, and m_2 is a mass hanging from the string. On the diagram, indicate the forces of gravity on both of these, marking the vector direction and the symbol for the force.

Step 2: Mass m_1 also experiences the normal force F_N and, when in motion, the force of friction f_{fr}. On the diagram, indicate these forces and label them.

Step 3: Mass m_2 is not subject to a normal force as m_1 is, but instead it is affected by the tension in the string. Indicate the tension *T* of the string as it is acting upon m_2.

Step 4: Mass m_1 is also connected to the same string and experiences the same tension *T*. The pulley serves to change the direction of the force, directing it horizontally. On the diagram, indicate the tension *T* of the string acting upon m_1.

Step 5: For simplicity, let's ignore the force of friction in this case. Note that the rope isn't an outside agent that can cause the whole system to speed up or slow down, but an object that simply holds the system together. So, we can ignore its effects on the net motion of the system of the cart and hanging mass. After this, the only external force acting on our cart/hanging mass system is gravity. Write the equation for this force in terms of the mass m_2 and the constant acceleration due to gravity.

Step 6: In calculating the acceleration of a system, be sure to take into account the mass of the entire system. In this case, the mass of the entire system includes both mass m_1 —including the mass of the cart as well as any additional load masses—and mass m_2. Write the equation for the acceleration of the system in terms of the masses m_1 and m_2, and the expression of the force due to gravity that you found in Step 5.

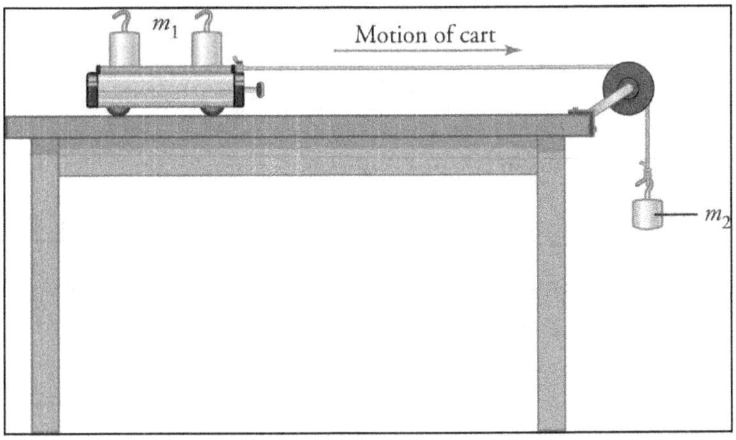

Figure 4.1: *The cart of mass* m_1 *accelerates along the table. The rate of acceleration is directly proportional to mass* m_2.

Assessments

1. What is acceleration? TEKS (4)(B)
2. How is the force of gravity on an object related to its mass? TEKS (4)(D)
3. Based on your examination of the free-body diagram, in which direction would you expect the cart to move? TEKS (4)(E)

Activity 2: Applying Constant Force TEKS (4)(A) TEKS (4)(B) TEKS (4)(D)

When a constant force is applied to an object or system, the object or system accelerates at a constant rate. If you know the applied force and mass of the system, you can calculate the acceleration and compare it to experimental results.

In this activity, you will set up the experiment shown in Figure 4.1. Using different masses for m_2, you will record the changes in velocity as the cart travels along the table. From this information, you will calculate the experimental acceleration.

Safety Precautions

- Be gentle with the electronics—sensors on the photogates must be handled carefully.
- Use the carts as directed; make sure the carts do not fall on the ground, collide with each other, or strike any one.
- Heavy weights can cause injury. Keeping overall system weight low is ideal. If you choose to experiment with heavier weights, ensure the area is clear.

For this activity you will need the following:

- Cart of known or measured mass
- Weights , a variety of them
- Pulley and string setup
- Meter stick
- Photogates
- Device to capture photogate data

For this activity you will work *in pairs* or *small groups*.

Procedure

Step 1: We are first going to measure speeds at two points using photogates, and then measure the time to go from one photogate to the other so we can experimentally determine the acceleration. Set up the cart and pulley system as shown in Figure 4.1, above. Lay the meter stick along the table, next to the string between the cart and the pulley. Set the cart at its starting position at the beginning of the meter stick.

Step 2: Arrange the two photogates somewhat apart from each other along the meter stick, as you did in the experiment on Acceleration. Be sure to set the first photogate (photogate A) far enough down the track so that it is not triggered by the cart at rest.

Step 3: Measure and record in Table 4.1 the distance from the starting position of the cart at rest to the first photogate. Also record the mass of the cart in Table 4.1.

Step 4: Secure selected mass m_2 in its designated position. Set the mass on the table next to the pulley until you are ready to execute the procedure. Record mass m_2 in the appropriate box in Table 4.2.

Step 5: Make sure the photogates are set to run in Gate mode.

Step 6: Position your cart at the beginning of the meter stick.

Step 7: Gently release mass m_2. Do not drop the mass from a height, since that would cause the cart to accelerate very suddenly, skewing your results. Also, always ensure that the string is taut before releasing the mass, so it does not hit the floor.

Step 8: Record in Table 4.2 the velocity of the cart at photogate A and at photogate B.

Step 9: Repeat Steps 6 through 8 twice more with the same mass, m_2, recording the data in Table 4.2. You should end up having three experimental values for each of the velocities at photogate A and photogate B.

Step 10: Now, set the photogates to run in Pulse mode. Keep the photogates set up as they were before. Now that you have the velocities at both photogates, you need to find the time it takes to travel between the photogates. This is the time it takes to go from the velocity traveled at photogate A to the velocity traveled at photogate B.

Step 11: Return the cart to its starting position at the beginning of the meter stick.

Step 12: Gently release mass m_2.

Step 13: Record in Table 4.2 the time it takes for the cart to pass through both photogates.

Step 14: Repeat Steps 11 through 13 twice more with the same mass m_2. You should end up having three experimental values for the time it takes to travel between photogate A and photogate B.

Step 15: Repeat Steps 4 through 14 twice more, with different masses for m_2. In all, at least three different masses should be used.

Step 16: Complete Table 4.3. The Trial 1 averages are calculated using Trials 1a, 1b, and 1c, and so on. Calculate the acceleration for each gate configuration using the averages.

Step 17: Compare the experimental accelerations. Are they close to each other? Why or why not?

Step 18: Using the graph paper in Figure 4.2 and the data you filled into Table 4.3, graph the average velocity vs. time plots for each of the three masses you used. The *x*-axis should be *time* and the *y*-axis should be *velocity*. The slope of each line represents the system acceleration. Use different colors for each of the three lines, and create a legend to show which lines represent which masses. What do you notice about each of the lines?

Table 4.1: Activity 2 Experimental Constants

Mass m_1 (mass of the cart) (g)	
Distance to the photogate A (cm)	

Table 4.2: Data Collected from the Photogates

	Mass m_2 (g)	Velocity at Photogate A (cm/s)	Velocity at Photogate B (cm/s)	Time Interval between Photogates A and B (s)
Trial 1a				
Trial 1b				
Trial 1c				
Trial 2a				
Trial 2b				
Trial 2c				
Trial 3a				
Trial 3b				
Trial 3c				

Table 4.3: Data Averages and Mathematical Analysis of Data

	Velocity at Photogate A (cm/s)	Velocity at Photogate B (cm/s)	Time Interval between Photogates A and B (s)	Acceleration between Photogates A and B (cm/s^2)
Trial 1 average				
Trial 2 average				
Trial 3 average				

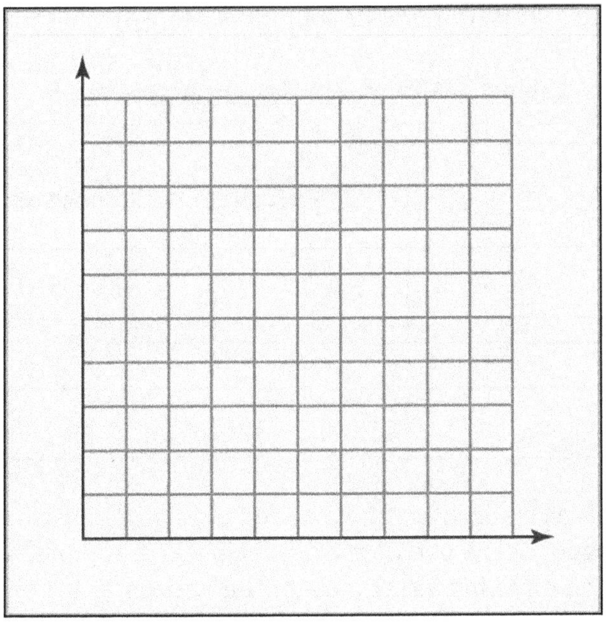

***Figure 4.2**: Graph paper*

Assessments

1. What is the source of the force causing the acceleration of the system used in Activity 2? TEKS (4)(B)
2. In what direction is the force being applied to the cart? TEKS (4)(B)
3. How did you change the acceleration of the cart? TEKS (4)(D)
4. When you increased the hanging mass, did that increase or decrease the acceleration of the cart? Why? TEKS (4)(B)
5. What measurements did you make that enabled you to calculate acceleration? TEKS (4)(D)
6. Using the equation from Step 6 of Part 1 of this lab, calculate the theoretical acceleration for each of the masses you used for m_2 in this part of the experiment. Fill in your answers in Table 4.4 below. Remember to use the mass of the cart as m_1. When applying the acceleration due to gravity, remember your units are $\mathrm{cm/s^2}$. TEKS (4)(D)

Table 4.4: Theoretical System Acceleration

Mass m_2 (g)	Acceleration of the System $(\mathrm{cm/s^2})$
1.	
2.	
3.	

7. Do the theoretical calculations match what you obtained experimentally? Why or why not?
8. Using the graph paper below and data from Table 4.4, graph the average velocity vs. time plots for each of the three masses you used. The horizontal axis should be *time* and the vertical axis should be *velocity*. Assume the velocity is equal to zero at time $t = 0$ for each mass, and draw each line using the calculated acceleration as the slope. Use different colors for each of the three lines, and create a legend to show which lines represent which masses. How are these plots the same or different from those you plotted from the experimental data? TEKS (4)(A)

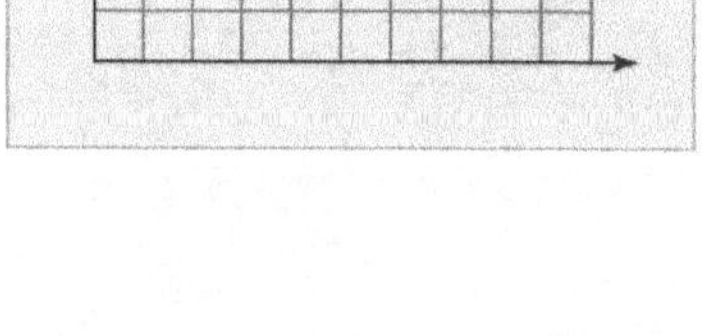

Activity 3: Effect of Force on Different Masses TEKS (4)(A) TEKS (4)(B) TEKS (4)(D)

Newton's second law tells us that the acceleration of an object or system is inversely proportional to the mass of the object or system. For a given applied force, the bigger the mass of the system, the lower the acceleration.

In this activity, you will explore the response of the system acceleration as the overall system increases its mass while still subject to the same force. Specifically, you will change the load mass on the cart while keeping the same hanging mass on the string.

Safety Precautions

- Be gentle with the electronics—sensors on the photogates must be handled carefully.
- Use the carts as directed; make sure the carts do not fall on the ground, collide with each other, or strike any one.
- Heavy weights can cause injury. Keeping overall system weight is ideal. If choosing to experiment with heavier weights, ensure the area is clear.

For this activity you will need the following:

- Cart of known or measured mass
- Weights, a variety of them
- Pulley and string setup—if no pulleys are available, dental floss over the edge of the table will work
- Meter stick
- Photogates
- Device to capture photogate data

For this activity you will work *in pairs* or *small groups*.

Procedure

Step 1: Set up the cart and pulley system as shown in Figure 4.1. Lay the meter stick along the table, next to the string between the cart and the pulley. Set the cart at its starting position at the beginning of the meter stick.

Step 2: Arrange the two photogates somewhat apart from each other along the meter stick, as you did in Part 2. Be sure to set the first photogate (photogate A) far enough down the track that it is not triggered by the cart at rest.

Step 3: Measure and record in Table 4.5 the distance from the starting position of the cart at rest to the first photogate. Secure selected mass m_2 in its designated position. Set the mass on the table next to the pulley until you are ready to execute the procedure. Record the mass of m_2 in Table 4.5.

Step 4: Record mass m_1—the cart plus load mass—in the appropriate box in Table 4.6.

Step 5: Make sure the photogates are set to run in Gate mode.

Step 6: Position your cart at the beginning of the meter stick.

Step 7: Gently release mass m_2. Do not drop the mass from a height, since that would cause the cart to accelerate very suddenly, skewing your results. Also, always ensure that the string is taut before releasing the mass, so it does not hit the floor.

Step 8: Record in Table 4.6 the velocity of the cart at photogate A and at photogate B.

Step 9: Repeat Steps 6 through 8 twice with the same mass m_1, recording the data in Table 4.6. You should end up having three experimental values for each of the velocities at photogate A and photogate B.

Step 10: Now, set the photogates to run in Pulse mode. Keep the photogates set up as they were before. Now that you have the velocities at both photogates, you need to find the time it takes to travel between the photogates. This is the time it takes to go from the velocity traveled at photogate A to the velocity traveled at photogate B.

Step 11: Return the cart to its starting position at the beginning of the meter stick.

Step 12: Gently release mass m_2.

Step 13: Record in Table 4.6 the time it takes for the cart to pass through both photogates.

Step 14: Repeat Steps 11 through 13 twice with the same mass m_1. You should end up having three experimental values for the time it takes to travel between photogate A and photogate B.

Step 15: Repeat Steps 4 through 14 twice more, with two different masses for m_1. In all, at least three different masses should be used.

Step 16: Complete Table 4.7. The Trial 1 averages are calculated using Trials 1a, 1b, and 1c, and so on. Calculate the acceleration for each gate configuration using the averages.

Step 17: Compare the experimental accelerations. Are they close to each other? Why or why not?

Step 18: Using the graph paper in Figure 4.2 and the data you filled into Table 4.7, graph the average velocity vs. time plots for each of the three masses you used. The horizontal axis should be *time* and the vertical axis should be *velocity*. The slope of each line represents the system acceleration. Use different colors for each of the three lines and create a legend to show which lines represent which masses. What do you notice about each of the lines?

Table 4.5: Activity 3 Experimental Constants

Mass m_2 (mass of the weight on the string) (g)	
Distance to the photogate A (cm)	

Table 4.6: Activity 3 Data Collected from the Photogates

	Mass m_1 (g)	Velocity at Photogate A (cm/s)	Velocity at Photogate B (cm/s)	Time Interval between Photogates A and B (s)
Trial 1a				
Trial 1b				
Trial 1c				
Trial 2a				
Trial 2b				
Trial 2c				
Trial 3a				
Trial 3b				
Trial 3c				

Table 4.7: Activity 3 Data Averages and Mathematical Analysis of Data

	Velocity at Photogate A (cm/s)	Velocity at Photogate B (cm/s)	Time Interval between Photogates A and B (s)	Acceleration between Photogates A and B (cm/s^2)
Trial 1 average				
Trial 2 average				
Trial 3 average				

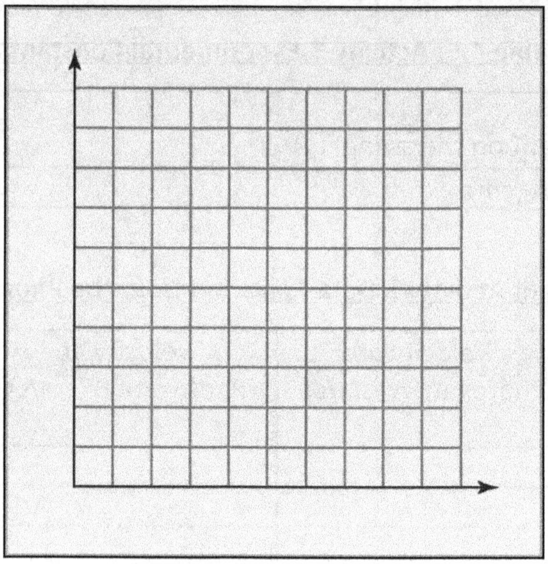

***Figure 4.2**: Graph paper*

Assessments

1. When you increased the mass on the cart, did that increase or decrease the acceleration of the cart? Why? TEKS (4)(B)
2. Using the equation from Step 6 of Part 1 of this lab, calculate your theoretical acceleration for each of the masses you used for m_1 in this part of the experiment. Fill in your answers in Table 4.8 below. Remember to use the mass of the hanging weight as m_2. When applying the acceleration due to gravity, remember your units are cm/s^2. TEKS (4)(D)

Table 4.8: Theoretical System Acceleration

Mass m_1 (g)	Acceleration of the System (cm/s^2)
1.	
2.	
3.	

3. Do the theoretical calculations match what you obtained experimentally? Why or why not?
4. Using the graph paper below and data from Table 4.8, graph the average velocity vs. time plots for each of the three masses you used. The *x*-axis should be *time* and the *y*-axis should be *velocity*. Assume the velocity is equal to zero at time $t = 0$ for each mass, and draw each line using the calculated acceleration as the slope. Use different colors for each of the three lines, and create a legend to show which lines represent which masses. How are these plots the same or different from those you plotted from the experimental data? TEKS (4)(A)

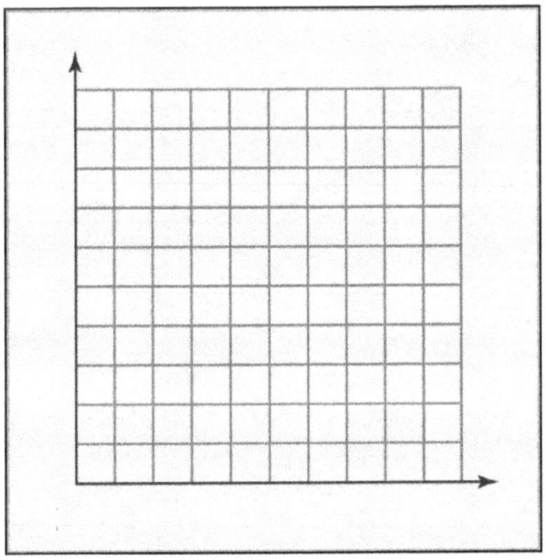

Lab 5: Motion in Two Dimensions

Learning about motion in *one* direction is a great way to learn about physics; however, objects in the real world do not just move back and forth in a straight line. Think about the route you take to school or to your favorite store. You most likely do not simply walk in a straight line in the direction of your destination. You likely have to turn corners and follow a more complicated path. In other words, your motion normally involves moving in *two* dimensions.

To describe motion in more than one dimension, we use vectors. **Vectors** are quantities that are fully described by both a magnitude and a direction. Vectors are graphically represented by arrows. The relative length of the arrow gives the magnitude while the arrowhead points in the direction of the motion. Some vector quantities include displacement, velocity, and acceleration. To determine the overall magnitude and direction of a traveling object, vectors can be added together to get a **resultant vector**. The resultant vector, also known simply as the resultant, combines the magnitudes and directions of all vectors of a given quantity into a single, overall magnitude and direction that describes the *net motion* of the object.

In this lab you will learn:

- how to add vectors in two dimensions graphically;
- how to determine the horizontal range of a projectile;
- how to use Hooke's law to calculate the spring constant of a spring.

Activity 1: Vector Addition TEKS (3)(F)

For objects traveling in just one dimension, the addition of vectors is simple. Just add together the magnitudes of the vectors, and the direction is determined by whether the resultant is positive or negative. When an object travels in two or more dimensions, it gets a bit more complicated. However, it does not matter in which order the vectors are placed when adding them together. The same resultant is found no matter the order in which you place the vectors.

One method for adding vectors is called the **tip-to-tail** method, or **head-to-tail** method. In this method, the vectors are placed tip-to-tail and the correct resultant vector goes from the *tail*—the end without the arrowhead—of the first vector to the *tip*—the arrowhead—of the final vector. You can then measure the resultant vector to determine its magnitude and direction.

Safety Precautions

- None

For this activity you will need the following:

- Protractor
- Ruler marked in cm
- Two pieces of graph paper
- Calculator
- Pencil
- Colored pencil

Procedure

Use the tip-to-tail method to add the following vectors.

Step 1: With your pencil, draw a vector arrow that is 8 cm long at 40 degrees above the positive horizontal axis ($+x$). Label your vector A.

Step 2: At the tip of the first vector, starting with the tail, draw another vector 10 cm long at 305 degrees from the positive horizontal axis. Label this vector B.

Step 3: At the tip of the second vector, starting with the tail, draw a third vector 6 cm long at 215 degrees from the horizontal. Label this vector C.

Step 4: Using a colored pencil, starting with the tail, draw the resultant vector by connecting the tail of vector A to the tip of vector C. The tip of the resultant vector should be touching the tip of vector C. Label the resultant R. See Figure 5.1, below.

Step 5: Measure the length of your resultant vector with a cm ruler and record the measurement in Table 5.1 for Trial 1.

Step 6: Measure with your protractor the angle with respect to the $+x$ axis of your resultant vector and record it in Table 5.1 for Trial 1.

Step 7: On the second sheet of graph paper, draw the same vectors in this order: C first, then A, then B.

Step 8: Using a colored pencil, starting with the tail, draw the resultant vector by connecting the tail of vector C to the tip of vector B. The tip of the resultant vector should be touching the tip of vector B. Label the resultant R.

Step 9: Measure the length of your resultant vector with your cm ruler and record it in Table 5.1 for Trial 2.

Step 10: Measure the angle of your resultant vector with your protractor and record it in Table 5.1 for Trial 2.

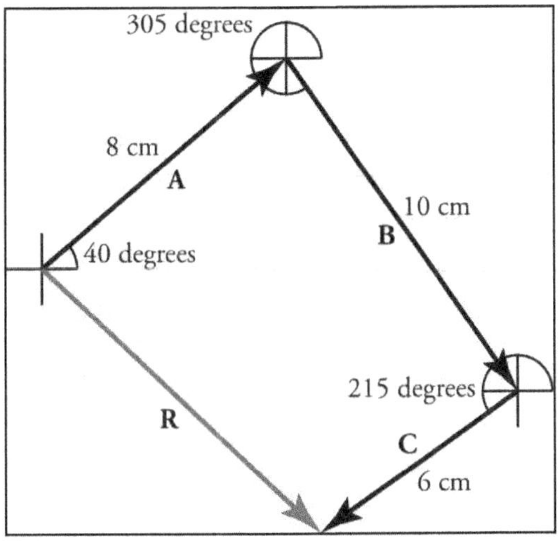

Figure 5.1: *Vectors with different magnitudes and directions can be added using the tip-to-tail method. The resultant vector gives the magnitude and overall direction of all of the vectors combined.*

Data and Observations

Table 5.1: Measuring Resultants

Trial Number	Length (cm)	Angle (degrees)
1		
2		

Assessments

1. A student chose to draw the same vectors drawn in this lab in the order B, C, A.
   a. Predict the magnitude and direction of this student's resultant vector.
   b. Explain your prediction. TEKS (3)(F)
2. Fill in the blank and then define the inserted term: The length of the arrow represents the ...... of the vector. TEKS (3)(F)
3. An additional vector D is added to the other vectors from Assessment 1. TEKS (3)(F)
   a. How would you insert a vector D into your drawing?
   b. How would this affect your resultant vector?

Activity 2: Projectiles TEKS (4)(C)

Projectiles are objects that are launched through the air and acted on only by the force of gravity. The curved path they follow is called a **parabola**. Projectiles have both *horizontal motion*—from their initial velocity—and *vertical motion*—from their initial velocity plus gravity pulling them down. The horizontal and vertical components act independently of one another, but act over the same amount of time. The acceleration along the horizontal and vertical components is also different. A projectile accelerates in the $-y$ -direction—due to gravity—but has no acceleration in the *x*-direction after it launches. Mathematical equations can be used to predict the **horizontal range**, which describes how far a projectile travels horizontally. Horizontal range is influenced by an object's initial velocity and height off the ground when fired.

Safety Precautions

- Set up your experiment in an open area so as not to hit anyone with a projectile.
- Remove any glass, cell phones, or other breakable items from the area.
- Stay away from the projectile as it is launched and until it hits the ground.

For this activity you will need the following:

- A table
- Two to three large books
- Ruler with a groove down the middle
- Masking tape
- Marble
- White paper
- Carbon paper
- Meter stick
- Calculator
- Stopwatch

For this activity you will work *in pairs*.

Procedure

Step 1: Lean the ruler against the stack of books to make a ramp, and secure the ruler to the books with masking tape. Follow the setup shown in Figure 5.2.

Step 2: Measure a 0.5-m distance from the edge of the table and mark this spot with a piece of masking tape. Then, place the bottom of the *ramp* there.

Step 3: Place the meter stick vertically against the edge of the table and place a piece of masking tape on the floor beneath the meter stick.

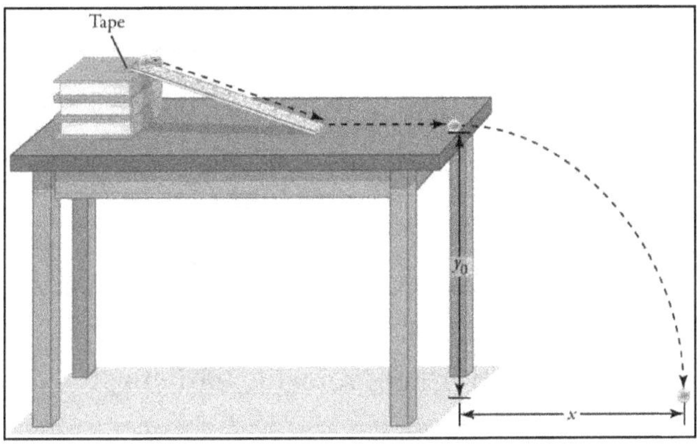

Figure 5.2*: Lean the ruler against the stack of books and secure it with tape to make the ramp. Place the ramp 0.5 m from the edge of the table.*

Step 4: Determine the initial velocity of the marble by timing how long it takes for the marble to travel from the bottom end of the ramp to the edge of the table. Then, divide the distance (0.5 m) by the time (in seconds). Be sure to stop the marble before it rolls off the table. Record your data in Table 5.2. Do this three times and find the average.

Step 5: Measure the height of the table and record it in Table 5.3 as y_0 , with the correct units.

Step 6: Calculate the total time the projectile is in the air (*t*) using the following equation

$$y = y_0 - \tfrac{1}{2}gt^2 .$$

You can use this equation because the projectile has no initial velocity in the vertical direction.

Record your value for *t* in Table 5.3.

Step 7: In Table 5.4, record v_x from Table 5.2 and *t* from Table 5.3.

Step 8: Solve for *x* using the equation $x = x_0 + v_x t$ and record the value in Table 5.4. This is your predicted horizontal range for your projectile.

Step 9: Place your white paper on the ground. The distance between the middle of your white paper and your masking tape line should equal your predicted horizontal range.

Step 10: Place the carbon paper carbon side down onto your piece of white paper.

Step 11: *Launch* your marble down the ramp and let it hit the floor.

Step 12: Lift up the carbon paper and measure the distance from the masking tape mark on the floor to the carbon mark on the paper.

Step 13: Compare your predicted horizontal range to the actual horizontal range.

Data and Observations

Table 5.2: Projectile Velocity

Trial	Distance (m)	Time (s)	Velocity (m/s)
1			
2			
3			
Average (v_x)			

Table 5.3: Other Projectile Variables

Variable	Value
y	0 m
y_0	
g	9.8 m/s²
t	

Table 5.4: Projectile Range

Variable	Value
x	
x_0	0 m
v_x	
t	

Assessments

1. Compare your expected horizontal range to your actual horizontal range. TEKS (4)(C)
    a. Were these numbers the same?
    b. If they were not the same, what are some possible sources of error?
2. Acceleration in the *y*-direction is due to .... How would this change if the projectile was on the moon? TEKS (4)(C)
3. Another book is added to the pile to make the ramp taller. TEKS (4)(C)
    a. How would this affect the initial velocity of the marble?
    b. How would this affect the horizontal range the marble would travel?

Activity 3: Simple Harmonic Motion TEKS (7)(A)

Simple harmonic motion is a type of motion in which a **displacement force**—the force that moves the object away from the resting position—has an equal and opposite **restoring force**—the force that brings the object back to the resting position. Examples of objects with simple harmonic motion include pendulums and springs. The **period** of an object in simple harmonic motion is the time it takes for one complete **oscillation**, or back and forth motion. For springs, the period depends upon a number called the spring constant, which is denoted by the letter *k* and has the units of newton-meters (N/m). Robert Hooke discovered a law, now called **Hooke's law**, which states that the restoring force of the spring is equal to the opposite of the spring constant times the length of the stretch of the spring $(F = -kx)$. If a spring is stretched farther than the restoring force can bring it back to its original shape, it has reached its **elastic limit**. After being stretched beyond its elastic limit, the spring will likely have stretched out coils and will not be able to store as much potential energy.

Safety Precautions

- Use caution when adding or removing weights from the spring.
- Do not stretch the spring past its elastic limit.
- Stay a safe distance away when starting the spring into motion and do not discharge springs in the direction of other people.
- Do not stretch strings near people's faces or eyes.

For this activity you will need the following:

- Table
- Ring stand with a ring
- Spring with a hook on the end
- Ruler
- Table clamp
- Hanging masses of various weights
- Calculator
- Stopwatch

For this activity you will work *in pairs*.

Procedure

Step 1: Place the ring stand, with the ring fastened near the top, at the edge of a table. Secure it firmly with a table clamp. Follow the setup shown in Figure 5.3.

Step 2: Attach the spring to the ring and add the lightest mass to the end. Record the mass, in kg, in Table 5.5.

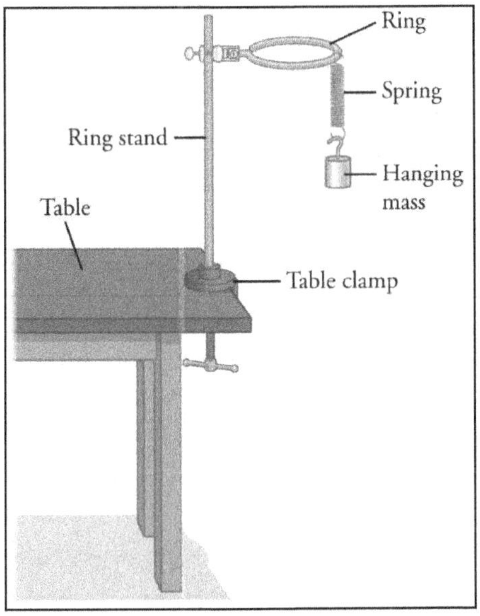

***Figure 5.3:** The ring stand should be clamped tightly to the edge of a table with the spring attached to the ring. Hanging masses can be suspended from the end of the spring.*

Step 3: Gently pull the mass and spring downward 3 cm. Carefully release the mass and spring and time how long it takes for 10 full **oscillations**, or up and down motions, to occur and record that time in Table 5.5.

Step 4: Divide the time for 10 oscillations by 10 to get the time for one period (T) and record the answer in Table 5.5.

Step 5: Repeat Steps 2–4 with different masses, being careful not to use so much mass that the spring stretches past its elastic limit.

Step 6: Calculate the spring constant (k) of the spring by using the mass and period for each trial and the following equation

$$T = 2\pi\sqrt{\frac{m}{k}}.$$

Step 7: Record your calculated spring constants in Table 5.5.

Data and Observations

Table 5.5: Oscillations of a Spring

Mass (kg)	Time for 10 Oscillations (s)	*T* (s)	*k* (N/m)

Assessments

1. The spring constant of the same spring should not change with each trial. TEKS (7)(A)
    a. Are all of your calculated spring constants the same?
    b. If not, explain why there may have been differences in your answers.
2. In which direction was the restoring force for the spring? TEKS (7)(A)
3. A more tightly wound spring that is harder to stretch is used for the same experiment. TEKS (7)(A)
    a. How would that affect the period of the spring? Explain.
    b. How would that affect the restoring force needed to bring the spring back to equilibrium? Explain.

Lab 6: Circular and Rotational Motion

A common misconception is that circular and rotational motions are the same. This is not true, however. **Circular motion** describes a mass moving in a circular path around a central point. **Rotational motion** occurs when a mass rotates on an **axis**, or an imaginary line through that body's center of mass. As shown in Figure 6.1, a Ferris wheel demonstrates both circular and rotational motion. The entire wheel rotates around the central axle, which demonstrates rotational motion. Circular motion describes the movement of an individual car around the circumference, or perimeter, of the wheel.

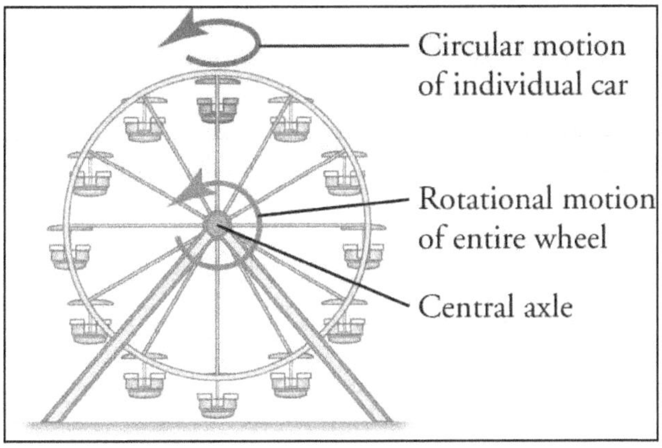

Figure 6.1: *A Ferris wheel exhibits both circular and rotational motion.*

<u>In this lab you will learn:</u>

- how to calculate the centripetal acceleration and force acting on an object in circular motion;
- how different masses affect centripetal force;
- how to determine what mass is required and the distance the mass should be from a fulcrum to balance a lever.

Activity 1: Uniform Circular Motion TEKS (4)(C)

Circular motion occurs when an object moves along the circumference of a circle, while keeping a constant radius from the center of the circle. **Uniform circular motion** is circular motion that occurs at a constant rate. **Tangential velocity** is the velocity of an object in circular motion—as opposed to linear velocity. Tangential velocity can be calculated with either the angular velocity (ϖ) or time. In this activity time is known, so we use the equation

$$\text{Tangential velocity} = v_t = \frac{2\pi r}{t}.$$

The unit for radius is meters (m), time is in seconds (s), and velocity is meters/second (m/s).

Centripetal acceleration of a body in circular motion is equal to the tangential velocity of the object squared, divided by the radius of the circle the object moves around.

$$\text{Centripetal acceleration} = \frac{(\text{tangential velocity})^2}{\text{radius}} \quad \text{or} \quad a_c = \frac{v^2}{r}$$

The units of centripetal acceleration are $\mathrm{m/s^2}$, where m is meters and s is seconds.

Centripetal force is the inward force that acts upon the object and keeps it moving around the circle. Newton's second law of motion states that centripetal force is equal to centripetal acceleration times the mass of the object.

In other words, centripetal force = mass x acceleration, or

$$F_c = ma_c = \frac{mv^2}{r}.$$

The units of centripetal force are newtons (N), where $1\ \mathrm{N} = 1\ \mathrm{kg \cdot m/s^2}$.

In this activity, you will collect data to determine the tangential velocity, centripetal acceleration, and centripetal force acting upon a mass in circular motion, as well as determine how different masses acting upon the stopper affect centripetal force.

Safety Precautions

- Make sure there is enough room to safely complete the lab.
- Do not spin a weight near any students or equipment.
- Make sure the stopper and weights are tied securely to the string.
- Inform your teacher of any broken glassware or any accidents.

For this activity you will need the following:

- Glass tubing
- Tape to cover tubing
- String
- Rubber stopper with holes
- Paper clips
- Washers
- Stopwatch

For this activity you will work *in pairs.*

Procedure

Step 1: Attach one end of the string to the rubber stopper.

Step 2: Wrap the glass tubing with tape to prevent breakage.

Step 3: Thread the second end of the string through the tubing.

Step 4: Place a paper clip on the string, as shown in Figure 6.2.

Step 5: Pull the string so the paper clip is near but not resting on the bottom of the tubing. Note the distance from the bottom of the tube and the clip.

Step 6: Keeping the string taut, measure the radius from the top of the tubing to the center of mass of the rubber stopper. Record this measurement in Table 6.1.

Step 7: Measure and record the mass of several washers in Table 6.1. Securely attach the washers to the end of the cord, below the glass tubing. Refer to Figure 6.2 to ensure the apparatus is properly set up.

Step 8: Holding the tubing vertically—perpendicular to the floor—have one partner spin the rubber stopper in a circle above his/her head. It is important to maintain a constant speed, and the paper clip must remain the same distance from the bottom of the tube. Practice this step a few times before moving on.

Step 9: Using the stopwatch, have the second partner record the amount of time to spin the stopper a full 20 revolutions. Record this time in Table 6.1.

Step 10: Repeat Steps 7, 8, and 9 two more times, being sure to change the number of washers and record their mass in Table 6.1 for each trial.

Step 11: Calculate tangential velocity, centripetal acceleration, and centripetal force acting upon the stopper and record them in Table 6.1. Use the mass of the washers to calculate centripetal force

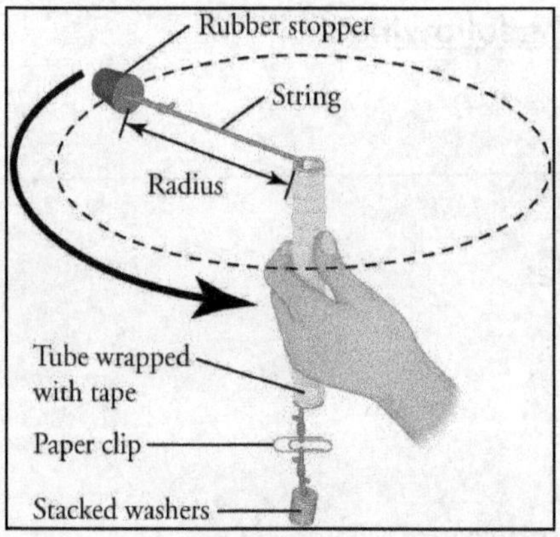

***Figure 6.2:** Apparatus setup for Activity 1, circular motion.*

Data and Observations

Table 6.1: Measurements from Rubber Stopper Experiment

Trial	Number of Revolutions	Total Time (s)	Radius (m)	Mass of Washers (g)	Tangential Velocity (m/s)	Centripetal Acceleration a_c (m/s²)	Centripetal Force F_c (N)
1	20						
2	20						
3	20						

Assessments

1. Jasmine is working on Activity 1 with her partner. While she is spinning the stopper overhead, the string breaks. On which path will the rubber stopper continue when the string breaks (a, b, c, or d in Figure 6.3)? Explain your choice using the concepts covered in this lab. Assume that the motion of the stopper continues in the plane in which it is spinning—that is, the stopper does not take off in either an upward or downward direction. TEKS (4)(C)

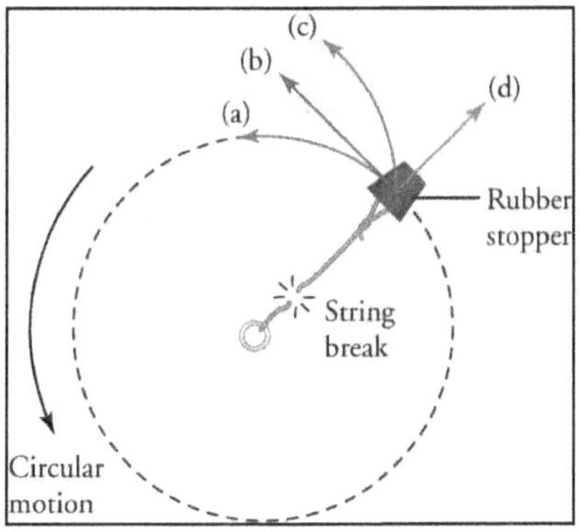

2. Explain how centripetal force changes as the mass of the object in motion, speed of the object, or radius of the circle increases. Why is it important to keep the paper clip at a constant distance from the bottom of the tubing? TEKS (4)(C)
3. What is the mass of a vehicle traveling at 15.0 m/s around a curve with a radius of 40.0 m if the centripetal force acting upon it is 3800 N? TEKS (4)(C)

Activity 2: Rotational Motion TEKS (4)(D)

Recall that circular and rotational motion are not the same: Rotational motion is the movement of a body around its axis or center of mass. A Ferris wheel spinning around its central hub shows rotational motion, as does Earth spinning on its axis. **Torque** measures the effect a given force exerts on rotational motion. Torque depends upon the magnitude of the applied force and the distance from the pivot point where that force is applied. The **pivot point** is the point about which a body rotates. A **fulcrum** is the supporting point of a lever on which it pivots. Many combinations of distance and force can result in the same torque. Figure 6.4 shows that less force is needed to loosen the nut as the distance between the force and the pivot point—the center of the nut—increases. However, the torque produced at both x_1 and x_2 will remain constant.

$$F_1 \cdot x_1 = F_2 \cdot x_2$$

In practical terms, you probably know this to be true: To move a really stubborn nut, you reach for a longer wrench so you can apply less force farther from the pivot point. You don't reach for a shorter wrench.

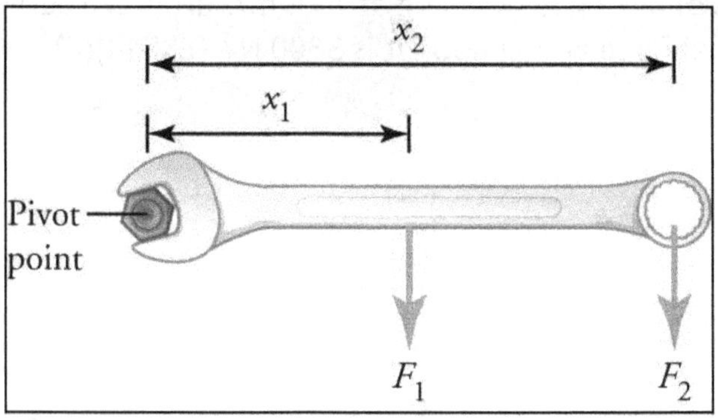

Figure 6.4: *You can produce the same torque to loosen the nut by applying a larger force* F_1 *at a shorter distance* x_1 *or a smaller force* F_2 *at a greater distance* x_2*. It is often more convenient to use a smaller force.*

Torque is calculated as

$$\tau = r_{\perp} F ,$$

where $r_{\perp}$ is the perpendicular distance from the pivot point to the line of the force, measured in meters, and *F* is the applied force. The units for torque are Nm (newton-meters).

Net torque is the sum of the combined torques acting upon a system. If a force moves a body in a counterclockwise motion, the torque is positive. If the force moves a body in a clockwise motion, the torque is negative. When the sum of all positive and negative torques is zero, the system is said to be in **static equilibrium** and will not rotate. Think of a mobile that is perfectly balanced, as in Figure 6.5. When the net torque of the hanging objects is zero, the mobile does not move—unless acted upon by an outside force. If the net torque is positive, the mobile turns counterclockwise; while if the net torque is negative, the mobile turns clockwise. Can you tell if the mobile in Figure 6.5 is moving or is in static equilibrium?

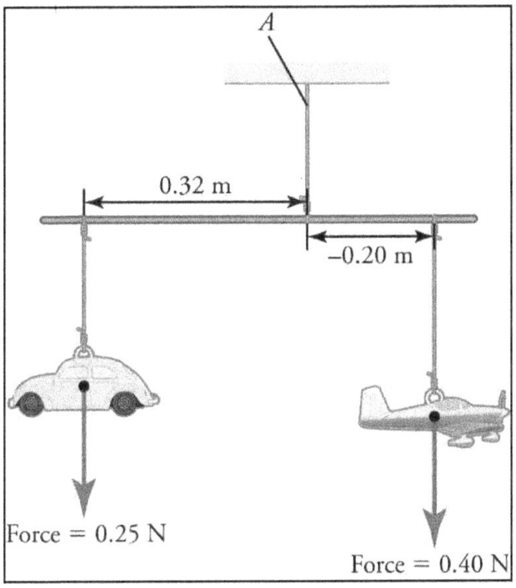

Figure 6.5: *The torque to the left of the pivot point (A) is* $0.08\ \text{N}\cdot\text{m}$ *and the torque to the right of the pivot point is* $-0.08\ \text{N}\cdot\text{m}$*. The sum of these two torques is zero; so, the mobile is in static equilibrium.*

In this activity, you will find the fulcrum of a meter stick and then arrange two weights on the balanced meter stick. You will need to determine the correct distance from the fulcrum to add a third weight such that the sum of all torques is zero and the meter stick is in static equilibrium. Typically, the force of a hanging mass—what is commonly referred to as weight—is equal to its mass times the acceleration due to gravity (*g*); however, in this activity, the gravitational acceleration can be canceled out as it occurs on both sides of the equation for net torque.

Safety Precautions

- Do not use the lab materials in any manner other than described in the lab.
- Use caution when handling the knife-edge.

For this activity you will need the following:

- Meter stick
- Knife edge
- Rubber bands
- Two 100-g weights
- Two 200-g weights
- One 300-g weight
- One 400-g weight

Procedure

Step 1: Find the center of mass of the meter stick by finding the point at which it balances on the knife-edge.
Step 2: Following the data in Table 6.2, hang masses the specified distances from the fulcrum by using rubber bands. The distances are given as negative to the right of the fulcrum and positive to the left of the fulcrum. Refer to Figure 6.6 to ensure the lab is set up correctly.
Step 3: Calculate the distance to the right of the fulcrum where the third mass should be hung, so the meter stick is in static equilibrium. Keep in mind the net torque should be zero.
Step 4: Record the calculated distance in Table 6.2, and then hang the third weight this distance to the right of the pivot point.
Step 5: Is the system balanced? If not, adjust the distance of the third weight until the meter stick is level. Record this distance in the table.
Step 6: Find the percent error between the calculated and observed distances and record the percent error in Table 6.2. The formula for percent error is

$$\text{Percent error} = \frac{|\text{Calculated - Observed}|}{\text{Calculated}} \times 100\%,$$

where the calculated value is the distance you calculated in Step 3 and the observed value is the distance the weight was actually hung to achieve static equilibrium, Step 5. Find percent error for all trials.

Step 7: Calculate the magnitude of the torque to the left of the fulcrum and record it in Table 6.2. This is *not* a net torque calculation; you will need to find the force exerted by the hanging weights for this calculation. Weight is a force that is the product of the mass of an object and the acceleration due to gravity $(g = 9.8\,\text{m/s}^2)$.

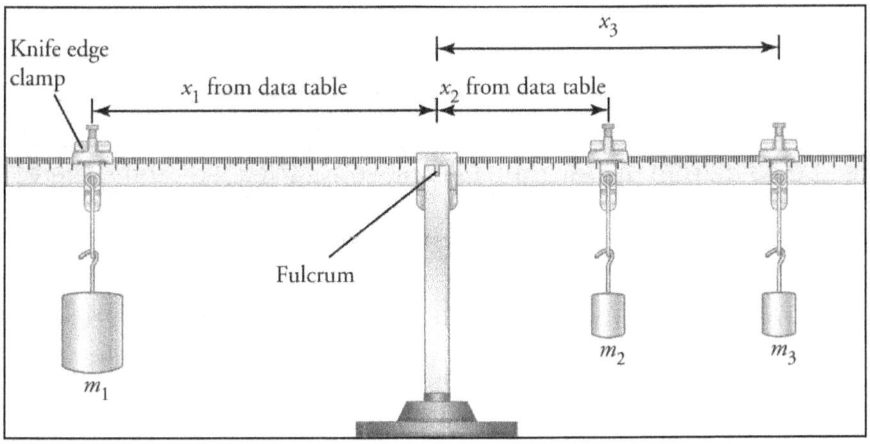

Figure 6.6: *Equipment setup for Activity 2, rotational motion.*

Data and Observations

Table 6.2: Measurements from Lever Experiment

Trial	Mass 1 in grams [m_1]	Distance in Meters to Left of Fulcrum [x_1]	Mass 2 in grams [m_2]	Distance in Meters to the Right of Fulcrum [x_2]	Mass 3 in grams [m_3]	Calculated Distance to Right of Fulcrum [x_3] (m)	Actual Distance (m)	% error	Torque to the Left of the Fulcrum
1	200.0	0.40	100.0	0.20	200.0				
2	300.0	0.50	100.0	0.60	200.0				
3	400.0	0.20	200.0	0.20	200.0				

Assessments

1. Do you think equal amounts of mass must hang on either side of the fulcrum to achieve balance? Calculate how much mass must be added to the left-hand side of a lever, at a distance of 0.40 m from the fulcrum, to achieve static equilibrium if there is a 275.0-g weight at distance $x_1 = -0.60\ \mathrm{m}$, to the right of the fulcrum. TEKS (4)(D)

2. Many torque applications need to take into consideration θ, the **angle of rotation**. The equation for torque can then be written as

$$\tau = F \cdot r \sin\theta \,,$$

   where *r* is the **lever arm**—perpendicular distance from pivot point to where the force is applied—and *F* is the applied force—mass of the object times acceleration due to gravity. Is this the same way net torque is calculated in this lab? If not, explain why certain quantities can be ignored in a net torque calculation. TEKS (4)(D)

3. The system in Figure 6.7 consists of two discs attached together at point A. The system is in static equilibrium when $r_1 m_1 = r_2 m_2$. Describe two ways that you can adjust the masses to make the system turn in a counterclockwise direction. TEKS (4)(D)

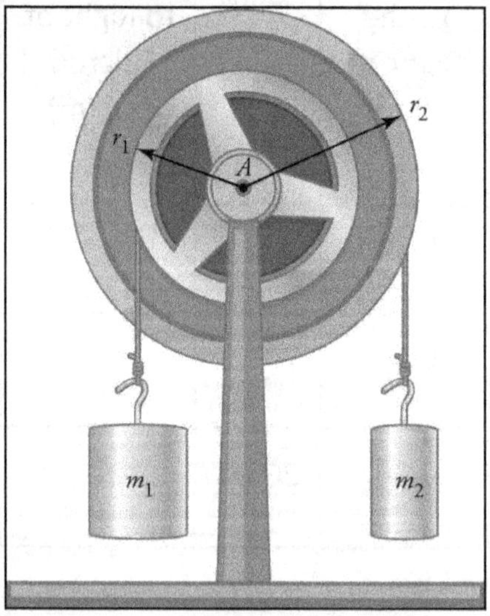

Figure 6.7: *Hanging balance system for Assessment 3.*

Lab 7: Work and Energy

Work is done when a force causes an object to move in the direction that the force is applied. When you lift a rock, work is done because a force generated by your arms causes the rock to move up, against the force of gravity. When you carry a rock on level ground, your arms do no work because they do not apply a force in the direction that the rock is moving. Work (*W*) is equal to the applied force (*f*) times the displacement (*d*) in the direction of the force, as shown in the equation

$$W = \mathbf{F} \cdot \mathbf{d} .$$

Work can only be done if energy is used. In fact, energy is sometimes defined as the ability to do work. Work and energy are closely related and both are measured with the same units. The SI unit for work and energy is the joule (J). One joule is equal to one newton-meter (N·m).

There are many different types of energy. **Kinetic energy** (*KE*) is energy due to motion. Kinetic energy depends on the mass of an object (*m*) and its speed (*v*),

$$KE = \frac{1}{2}mv^2 .$$

Energy can also be stored. Stored energy is called **potential energy** (*PE*) because it has the potential to do work**.** Gravitational potential energy is potential energy due to the position of an object above Earth's surface. For example, water behind a dam has gravitational potential energy that can be used to generate electricity. As the water falls into a turbine, its potential energy is converted to kinetic energy, which is then converted to electrical energy by the turbine.

Energy is conserved, which means that if one type of energy increases, another type of energy must decrease. Consider what happens as you bike up a long, steep hill; chemical potential energy from the food you have eaten is converted to gravitational potential energy. As you coast back down, gravitational potential energy is converted to kinetic energy as you accelerate.

To change an object's gravitational potential energy (PE_g), we need to apply a force. The change in gravitational potential energy that occurs equals the object's weight (*w*) times its change in height (*h*). Weight, in turn, is equal to the object's mass times the acceleration due to gravity (*g*). Therefore, we can calculate an object's gravitational potential energy by using the equation

$$PE_g = wh = mgh .$$

In this lab, you will explore the relationship between work and energy. In Activity 1, you will measure the change in kinetic and potential energy of a marble as it rolls down a ramp. In Activity 2, you will use a simple machine called a lever to explore the relationship between work, force, and distance. In Activity 3, you will run up a staircase and calculate how much work you do per unit time. Work done per unit time is called **power**.

<u>In this lab you will learn:</u>

- the nature of kinetic-energy and potential-energy transformations;
- how a simple machine is able to reduce force while conserving energy;
- how to increase power without increasing the amount of work done.

Activity 1: Conservation of Energy TEKS (6)(B), TEKS (6)(D)

In a mechanical system, potential energy is converted to kinetic energy, and kinetic energy is converted to potential energy. In a mechanical system, **mechanical energy** (*ME*) is the sum of potential energy and kinetic energy, as indicated by the equation

$$ME = KE + PE\ .$$

In Figure 7.1, all of the bicycle's potential energy is converted to kinetic energy as the bicycle rolls down the hill, and so the total mechanical energy of the system remains constant. In real life, however, some energy is always lost as heat. For example, as the bicycle's tires roll on the road, friction converts some of the bicycle's kinetic energy into heat, resulting in a reduction in mechanical energy in the system.

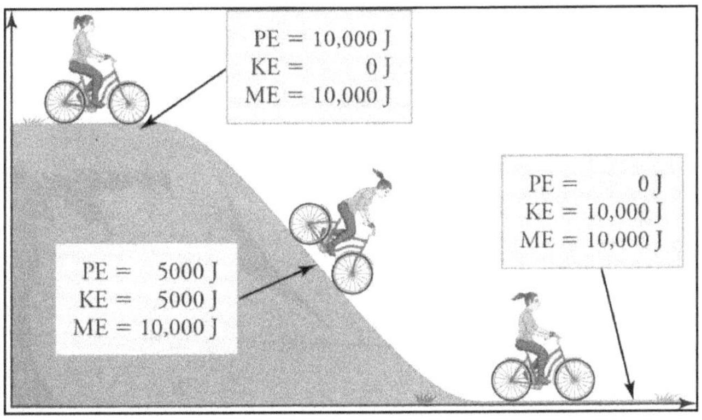

Figure 7.1: *As a bike coasts downhill, gravitational potential energy is converted to kinetic energy. Mechanical energy is the sum of these two forms of energy.*

Efficiency is a measure of how much energy is retained as mechanical energy. As efficiency increases, so does the work that is done by the system. Efficiency is usually calculated by dividing the work output (W_O) of a system by the work input (W_I), where work input is the initial amount of mechanical energy and work output is the amount of mechanical energy retained after work is done. Efficiency is typically reported as a percentage, written as

$$\%\ \text{Efficiency} = \frac{W_O}{W_I} \times 100\%\ .$$

In this lab you will build a ramp and calculate the potential energy, kinetic energy, and mechanical energy of a marble at various points as it rolls down the ramp. From your calculations, you will determine the efficiency of this mechanical system.

Safety Precautions

- Be careful not to leave marbles lying on the floor where people might step on them and slip.

For this activity you will need the following:

- Two meter sticks
- 2-inch-wide masking tape
- Small binder clip
- 10 identical marbles
- Level
- Stopwatch, precise to 0.01 seconds
- Marking pen
- Ruler
- Balance
- String

For this activity you will work *in pairs.*

Procedure

Step 1: Assemble the ramp.

- Place a meter stick upside-down on the lab bench so that the zero mark is at the top right. Align it so that the length of the stick is flush with the edge of the bench.
- Press the second meter stick against the side of the bench with the numbers facing out, and align the top long edge of this stick with the back long edge of the other (Figure 7.2). Fit the two sticks together so there is no gap and tape them along the entire length (one partner can hold the sticks while the other one tapes). The two meter sticks should be taped together along their long sides to form a V-shape.
- Set the flat edge of the binder clip on the bench and make a V with the wire handles. Carefully set the taped meter sticks inside this V with the taped edge pointing down. The zero mark should be close to the binder clip, and the other end of the meter sticks should rest on the bench.
- To test your ramp, place a marble at the zero mark and let it go. Does the marble roll smoothly and appear to pick up speed as it rolls? Does it roll too fast for you to measure its time of travel with a stopwatch? If not, adjust the tape and the meter sticks until you get a smooth roll of the marble.

Note that the marble rolls, which gives it rotational kinetic energy. However, for the purposes of this experiment, we will ignore this component of its energy. Although please note that rotational kinetic energy contributes a significant amount to the total kinetic energy of the rolling sphere.

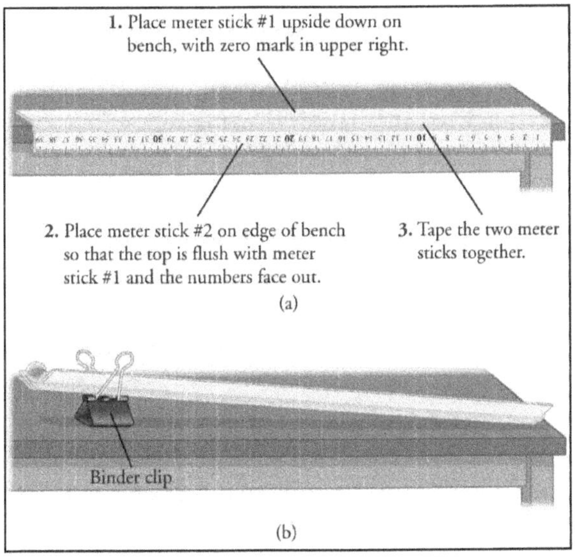

Figure 7.2: *(a) Tape two meter sticks together to form a ramp. (b) Rest one end of the ramp on a binder clip and the other end on the lab bench.*

Step 2: Mark the ramp with tape to make it easier to visualize where to start and stop the stopwatch. Cut four 10-cm-long pieces of tape. Place a piece of tape along each of the following ranges on the ramp:

0–10 cm

10–20 cm

40–50 cm

90–100 cm

The tape should be placed high enough that it does not interfere with the rolling marble.

Step 3: Determine the ramp height at each measurement location.

- Cut a piece of string a bit longer than 1 m.
- Hold the string horizontally with one end at the zero mark in the trough of the ramp and the other end above the 1 m mark. Use the level to ensure that the string is horizontal.
- The partner not holding the string should measure the vertical distance from the string to the trough at 10 cm, 20 cm, 50 cm, and 100 cm. Record each displacement in Table 7.1. Calculate the height above the 100 cm mark in centimeters and meters and record your results in the table.

Step 4: Measure the speed of the marble at various locations.

For each measurement, one lab partner should set the marble at the zero point of the ramp and then let it go. The other lab partner should measure the speed at a given point by starting the stopwatch as the marble passes the beginning of a piece of marking tape and stopping the stopwatch as the marble reaches the end. The following are a few tips about time measurement:

- You will need to restart the marble at the top of the ramp for each measurement.
- Don't push the marble; this will boost its kinetic energy.
- Getting a good measurement can be tricky; practice a few times before you begin recording your data.
- The marble is continuously accelerating, which means that the velocity is continuously increasing. Thus, your velocity measurement is a rough approximation.

Once you are done practicing, make three time measurements for each taped location on the ramp. Record the data in Table 7.2 and calculate the average time for each position. Next, calculate the speed as shown in the table.

Step 5: Measure the mass of the marble.

Measure the mass of 10 marbles in grams. Then use the following equation to calculate the mass of 1 marble in kilograms:

$$\text{mass of one marble (kg)} = \frac{\text{mass of 10 marbles (g)}}{10} \times \frac{0.001\,\text{kg}}{1\,\text{g}} .$$

Record the mass of one marble in kilograms in Table 7.3 and Table 7.4, as indicated. You may assume that the mass of each marble is the same in all cases.

Step 6: Calculate gravitational potential energy.

Copy the heights in meters that you calculated in Table.7.1 to Table 7.3. Calculate potential energy as indicated in Table 7.3.

Step 7: Calculate kinetic energy.

Copy the average speeds from Table 7.2 to Table 7.4. Calculate kinetic energy as indicated in the table.

Step 8: Calculate mechanical energy and percent efficiency.

Copy the calculated gravitational potential energies and kinetic energies to Table 7.5. Calculate the mechanical energy and the percent efficiency.

Data and Observations

Table 7.1: Height of Ramp

Position on Ramp (cm)	Displacement from String to Trough (cm)	Height (cm)*	Height (m)
0 (beginning of ramp)	0		
10			
20			
50			
100 (end of ramp)		0	0

*height in cm = (displacement at end of ramp) – (displacement at position *x*)

Table 7.2: Velocity Measurements

Position on Ramp (cm)*	Time (s)				Velocity (m/s)**
	Trial 1	Trial 2	Trial 3	Average	
0 (beginning of ramp)					
10					
20					
50					
100 (end of ramp)					

*Note that the time trials are made for 0-10 cm, 10-20 cm, 40-50 cm, and 90-100 cm. However, the marble is continuously accelerating, so the calculated velocity is an average of that at 10 cm, 20 cm, 50 cm, and 100 cm.

**The marble traveled 10 cm, or 0.1 m, in each trial. Thus, velocity is calculated by dividing 0.10 m by the average time (*t*) in seconds ($v = \frac{0.10 \text{ m}}{t}$).

Table 7.3: Potential Energy

Position	Marble Mass (kg)	Acceleration Due to Gravity (m/s²)	Height (m)	Gravitational Potential Energy (J)*
0 cm (beginning of ramp)		9.8		
10 cm		9.8		
20 cm		9.8		
50 cm		9.8		
100 cm (end of ramp)		9.8	0	

*Gravitational potential energy is the product of mass, acceleration due to gravity, and height ($PE_g = mgh$).

Table 7.4: Kinetic Energy

Position	Mass (kg)	Velocity (v) (m/s)	v^2	Kinetic Energy (J)*
0 cm				
0-10 cm				
10-20 cm				
40-50 cm				
90-100 cm				

*Kinetic energy is equal to one-half the mass times the velocity squared ($KE = \frac{1}{2}mv^2$).

Table 7.5: Mechanical Energy

Position (cm)	PE_g (J)	KE (J)	Mechanical Energy (J)*	% Efficiency**
0				-
10				
20				
50				
100				

*Mechanical energy is the sum of potential energy and kinetic energy ($ME = KE + PE_g$).

**Percent efficiency is equal to the mechanical energy at a given position divided by the initial amount of mechanical energy, multiplied by 100 percent. At the top of the ramp, the marble has only gravitational potential energy, so the initial mechanical energy is equal to the gravitational potential energy at the top of the ramp. Therefore, efficiency can be calculated with the equation

$$\%\ \text{Efficiency} = \frac{ME(\text{position } x)}{PE_{\text{g,top of ramp}}} \times 100\% .$$

Assessments

1. Jack and Jill climb a hill that is 200 m tall. TEKS (6)(B), TEKS (6)(D)
   a. If Jack has a mass of 65 kg, how many joules of gravitational potential energy does he gain on the climb?
   b. Jill weighs less than Jack. How will her change in gravitational potential energy compare to Jack's? Why?
2. A student determines that a block has 100 J of gravitational potential energy at the top of a ramp. He lets the block slide down the ramp and determines that it has 87 J of kinetic energy at the bottom of the ramp. How much mechanical energy is lost? TEKS (6)(B), TEKS (6)(D)

Activity 2: Levers (TEKS 6D)

A simple machine is a device that makes work easier by reducing the force that must be applied. Recall that work equals force times distance, shown as

$$W = \boldsymbol{F} \cdot \boldsymbol{d} \ .$$

Work and energy are conserved, which means that a simple machine cannot increase the amount of work that is done. Instead, the machine decreases force by increasing the distance over which the force is applied.

Consider a simple machine called a **lever**, which consists of a stiff rod and a pivot point called a fulcrum. In Figure 7.3, the right side of the lever is weighed down by a block. To balance this load, a downward input force (f_i) must be applied to the left side of the fulcrum. The input force induces an output force (f_o) that causes the block to move up by a height (h_o).

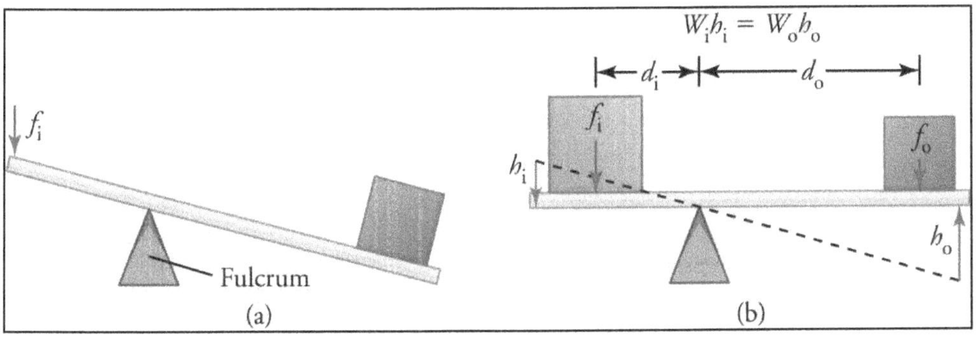

Figure 7.3: *A. An input force (f_i) must be applied to the left side of the fulcrum to lift the block on the right side. B. The lever reduces the force that must be used to lift a load by increasing the distance over which the force is applied.*

Recall that work is equal to the magnitude of the force times the distance over which the force is applied. Energy is conserved, so the maximum theoretical work output of a machine is equal to the work input

$$W_o = W_i \text{ , therefore } f_o h_o = f_i h_i \ .$$

Notice in Figure 7.3 that, because the fulcrum is off-center, the input load moves a smaller distance than the output load. As a result, the force required to move the output load is increased. A device that reduces the force that must be applied to move a load is called a simple machine. The factor by which a simple machine reduces the force is called the **mechanical advantage** (*MA*).

The change in height of the block (*h*) is proportional to the distance from the fulcrum, *d*. Thus, mechanical advantage can be calculated by dividing the distance of the input load from the fulcrum (d_o) by the distance of the output load from the fulcrum (d_i),

$$MA = \frac{f_o}{f_i} = \frac{d_i}{d_o}.$$

In this activity, you will build a lever and observe the relationship between work, force, and distance. From your observations, you will assess how the mechanical advantage of the lever depends on the position of the fulcrum.

Safety Precautions

- Be careful not to leave pennies lying on the floor where people might step on them and slip.

For this activity you will need the following:

- Large binder clip
- Ruler
- Pennies

For this activity, you will work *in pairs*.

Procedure

Step 1: Assemble the lever.

- Place a binder clip on its side on the lab bench. The handles should be extended so that they lie flat on the bench. This will be the fulcrum of your lever.
- Carefully balance a ruler on top of the fulcrum as shown in Figure 7.4.
- While holding the ruler, place a stack of two pennies 6 cm to the left of the fulcrum. Place another stack of two pennies 6 cm to the right of the fulcrum. Carefully adjust the ruler until it is balanced.
- The load on the left is the resistance load (f_r).
- The load on the right is the effort load (f_e).

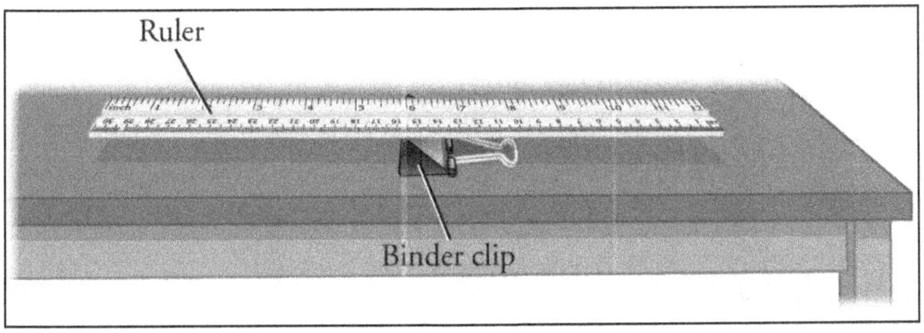

Figure 7.4: *Rest a ruler on a binder clamp to form a lever.*

Step 2: Change the resistance load and move the effort load to balance the lever.

- Change the number of pennies on the left (the resistance load) as indicated in Table 7.6, but always keep the distance of this load at 6 cm. Each time you change the number of pennies in the resistance load, move the pennies on the right (the effort load) until the lever is balanced. Be sure to grip the ruler each time you move pennies; you will need to carefully adjust the position of the pennies to get the lever to balance.
- Measure the distance between the pennies on the right and the fulcrum; this is the effort distance (d_e). Record your results in Table 7.6.

Step 3: Calculate the work done on each side of the fulcrum by multiplying the number of pennies (f) by the distance (d) and record your results in Table 7.6.

Step 4: Calculate the mechanical advantage as described in Table 7.6. Record your results in the table.

Data and Observations

Table 7.6: Mechanical Advantage

Resistance Load (left)			Effort Load (right)			Mechanical Advantage	
f_r (number of pennies)*	d_r (cm)	W_r $(f_r \times d_r)$	f_e (number of pennies)*	d_e (cm)	W_e $(f_e \times d_e)$	$\frac{f_r}{f_e}$	$\frac{d_e}{d_r}$
1	3		1	3			
2	3		1				
4	3		1				

*Both f_r and f_e are measured in units of pennies.

Assessments

1. A man wants to lift a 100 N load but can only apply 10 N of force. He uses a lever in which the fulcrum is 0.1 m from the resistance load. How far from the fulcrum must the effort load be to balance the lever? TEKS (6)(D)

2. As the distance of the effort load from the fulcrum increases, what happens to the mechanical advantage? Why? TEKS (6)(D)

Activity 3: Human *Horse* Power TEKS (6)(C)

Recall that gravitational potential energy is energy that an object possesses due to its height above Earth's surface. The relationship between the work that can be done and the gravitational potential energy used is

$$PE_g = mgh = W = fd\ .$$

It always takes the same amount of force to lift a load a given distance. However, the time it takes to lift the load may vary. **Power** (*P*) is a measure of the amount of work done in a given amount of time,

$$P = \frac{W}{t}\ .$$

Historically, much work was done by using horses. For this reason, power was often measured in units of horsepower. One horsepower equals 550 foot-pounds per second, and this is roughly the amount of power a horse can produce. Today, power is more commonly measured in **watts** (W). A watt is equal to one joule per second.

<u>Safety Precautions</u>

- Use care when running up the staircase.

<u>For this activity you will need the following</u>:

- Scale
- Ruler
- Stopwatch
- Staircase

For this activity you will work *in pairs*.

Procedure

Step 1: Measure your mass in kilograms. Record your mass and your partner's mass in Table.7.7.

Step 2: Measure the height of one step and multiply by the number of steps in the staircase. Record the total height of the climb, in meters, in Table 7.7.

Step 3: Use a stopwatch to measure the amount of time, in seconds, it takes you to run up the steps. Record the time in Table 7.7.

Step 4: Calculate your increase in gravitational potential energy from the bottom to the top of the staircase.

Step 5: Calculate the power you produced.

Data and Observations

Table 7.7: Power

Name	Mass (kg)	Total Height of Steps (m)	Gravitational Constant (m/s^2)	Time (s)	Increase in PE_g (J)*	Power (W)**
			9.8			
			9.8			

* $PE_g = mgh$

**The work you do in climbing the stairs is equal to your gain in potential energy.

Therefore, $P = \frac{PE_g}{t}$.

Assessments

1. Bob has a mass of 80 kg and Linda has a mass of 40 kg.
    a. How much work will Bob and Linda each do if they climb a staircase 3 m tall?
    b. Is it possible for Bob and Linda to produce the same amount of power while climbing the staircase? How? TEKS (6)(C)

Lab 8: Linear Momentum

Momentum is an important quantity in physics. It is a derived quantity of motion calculated from the measured quantities of mass and velocity. In one-dimensional motion, linear momentum, ***p***, is a vector with the same direction as velocity and magnitude proportional to that of the velocity vector. Momentum is measured in units of kg·m/s and is expressed mathematically as,

$$p = mv.$$

Collisions are forceful interactions between objects that involve an exchange of both kinetic energy and momentum. Would you prefer to collide with an object that is small or large? Would you prefer to collide with an object that is moving towards you or away from you? Considering collisions involving bouncing balls and peaches, are all collisions alike? In this lab, we will study kinetic energy and momentum in different types of one-dimensional collisions.

<u>In this lab you will learn:</u>

- how to distinguish between elastic and inelastic collisions;
- how to calculate the momentum of a physical system;
- how to solve collision problems by applying the law of conservation of momentum.

Activity 1: Elastic Collisions TEKS (6)(D)

When two balls collide, they bounce off each other. If kinetic energy is conserved, we define this interaction as an **elastic collision**. In this activity, you will be measuring the velocities of vehicles undergoing elastic collisions on a nearly frictionless track. You will calculate both kinetic energy and momentum before and after the collision to see if these are conserved in the collision. You will then change the mass and the velocity of the vehicles to observe how these variables affect the conservation of kinetic energy and momentum.

Safety Precautions

- Place the cars gently onto the track from above to avoid scratching the surface.
- Do not move the cars horizontally unless the compressed air supply is turned on.
- Wear safety goggles to prevent eye injuries from particles released during collisions.
- Do not throw or propel cars off the track or toward other students.

For this activity you will need the following:

- One frictionless compressed air track with two encoder strips along the length of the track, or a smooth track-like surface (such as a dry-erase board with two parallel meter sticks serving as the edges of a track). A commercially available air track is the Vernier Dynamics Cart and Track System with Motion Encoder (Order Code DTS-EC) See: http://www.vernier.com/files/manuals/dts-ec.pdf
- Two dynamics cars (with the same initial mass and removable weights to attach to cars). Cars are equipped with optical encoder and infrared (IR) transmitters.
- Two IR receivers attached to each end of the track.
- Additional items needed to improvise a track:
    - One roll of masking tape to hold meter sticks in place on track surface
    - Four digital timers precise to one tenth or one hundredth of a second
    - Two-meter-long smooth boards for incline ramps
    - Two meter sticks marked in cm
    - One cardboard box or stack of books to elevate ramp

For this activity you will work *in groups of four* (two to handle the cars and two to run the data collection from the sensors)

Procedure

Step 1: Set up the air track according to your teacher's directions. For the Vernier equipment, there are two encoder strips, one on each side of the track, with the receiver on the same side as the encoder strip for the car closest to it. The encoder strip consists of alternating black and white bars with a 4 mm period, allowing the optical sensor to detect the passage of the bars as the car moves. No alignments or adjustments are necessary, as one receiver attaches firmly to each end of the track. Consider reversing the direction of one receiver so that the same direction is positive for each system. Put the cars together, and zero both systems. This will put the cars on the same coordinate system; if they move together in contact, their position readings will be the same. Connect the receiver to an interface. If using a computer, connect the interface to your computer and launch the software.

Step 2: Turn on the car by pressing the power button. The car will glow blue when power is on. Place the car down vertically onto the track, with the wheels in the grooves, and with the blue light facing the receiver.

Step 3: Weigh the cars on a balance and record the masses in Table 8.1.

Step 4: Launch the cars near the center of the air track. Give one car a push, directing it away from the other car. Start with a slow, gentle push for best results. The cars will bounce off the end of the track and rebound, reaching a stable speed before colliding with each other (Figure 8.1). Be sure to stay out of the way of the motion sensors. Continue taking data as the cars bounce off one another. Then hit the *Stop* button, using the cursor. With two sensors appropriately placed on the underside of the cars, a narrow infrared beam transmits motion data from the car sensors to the nearest receiver and can detect a change in position with 1 mm resolution, as well as the direction of travel.

Step 5: The computer will calculate and plot the change in velocity and travel direction of each car before and after the collision. Record the velocity of each car before they collide as the initial velocity, and after they collide as the final velocity, in Table 8.1.

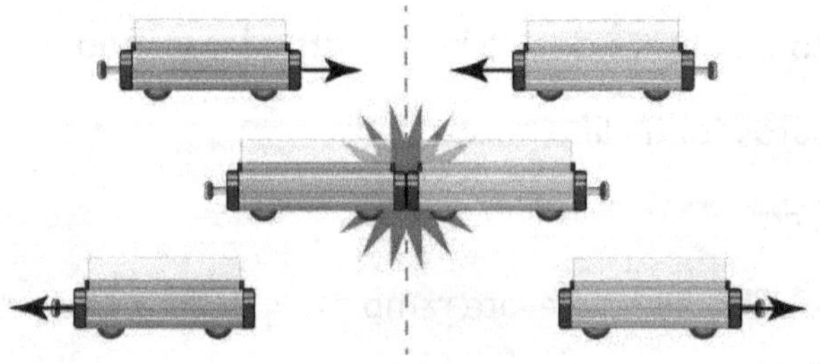

Figure 8.1: *In the three stages of an elastic collision, cars approach, collide, and then move away from each other without loss of energy. In the collision step, all of the kinetic energy is converted to potential energy in the spring. In Step 3, the potential energy has been reconverted to kinetic energy.*

Step 6: Repeat Steps 2-5, changing the initial velocity of one of the cars by giving the car a harder initial push. The car should move approximately twice as fast as the car moved in Trial 1.

The actual velocity is not important since the computer will calculate the velocity. Record the initial velocity as Trial 2 in Table 8.1.

Step 7: Repeat Steps 2-5 with one car having a different mass (with the addition of up to four 125 g weights). Record the initial velocity as Trial 3 in Table 8.1.

Step 8: Calculate initial and final kinetic energy for each car $(KE = \frac{1}{2}mv^2)$, using the velocities immediately before and after the collision. Record in Table 8.2.

Step 9: Calculate initial and final momentum (**p**=mv) for each car using the velocities immediately before and after the collision. Record in Table 8.2.

Step 10: Record in Table 8.3 the total kinetic energy before and after colliding, using the equations:

$$\text{Initial total KE} = \tfrac{1}{2}m_1v_{1i}^2 + \tfrac{1}{2}m_2v_{2i}^2$$
$$\text{Final total KE} = \tfrac{1}{2}m_1v_{1f}^2 + \tfrac{1}{2}m_2v_{2f}^2 \; ,$$

where m_1 and v_1 refer to the mass and velocity of one object while m_2 and v_2 refer to the mass and velocity of the other object. The letter *i* refers to initial and *f* refers to final. As the two vehicles are travelling in opposite directions, you will need to apply a negative sign to velocity on the left side of the equation and a positive sign to velocity on the right side of the equation.

Step 11: Compare the total kinetic energy before and after the elastic collisions by recording the ratio of final to initial total kinetic energy, and indicate in Table 8.3 whether kinetic energy is conserved.

Step 12: Record in Table 8.4 the total momentum before and after the collision, using the equations

$$\text{Initial momentum} = m_1v_{1i} + m_2v_{2i}$$
$$\text{Final momentum} = m_1v_{1f} + m_2v_{2f} \; .$$

Step 13: Compare the total momentum before and after the elastic collisions by recording the ratio of final to initial total momentum, and record in Table 8.4 whether momentum is conserved.

Step 14: Compare Trial 1 to Trial 2 in all experiments to determine whether changing the velocity of a car changes the conservation of energy, and record in Table 8.3. Also determine whether changing velocity changes the conservation of momentum, and record the result in Table 8.4.

Step 15: Compare Trial 1 to Trial 3 in all experiments to determine whether changing the mass of a car changes the conservation of energy, and record in Table 8.3. Also determine whether changing mass changes the conservation of momentum, and record the result in Table 8.4.

Data and Observations

Table 8.1 Elastic Collision: Mass and Velocity Measurement

Trial Run and Car Number	Mass of Car (kg)	Initial Velocity (m/s)	Final Velocity (m/s)
1 - Car A			
1 - Car B			
2 - Car A			
2 - Car B			
3 - Car A			
3 - Car B plus 500 g			

Table 8.2 Elastic Collision: Kinetic Energy and Momentum

Trial Run and Car Number	Initial Kinetic Energy (J)	Final Kinetic Energy (J)	Initial Momentum (kg m/s)	Final Momentum (kg m/s)
1 - Car A				
1 - Car B				
2 - Car A				
2 - Car B				
3 - Car A				
3 – Car B plus 500 g				

Table 8.3 Conservation of Kinetic Energy in Elastic Collisions

Trial Number	Total Initial Kinetic Energy (J)	Total Final Kinetic Energy (J)	Ratio of Final KE to Initial KE	Kinetic Energy Is Conserved?
1				
2				
3				

Table 8.4 Conservation of Momentum in Elastic Collisions

Trial Number	Total Initial Momentum $(\mathrm{kg \cdot m/s})$	Total Final Momentum $(\mathrm{kg \cdot m/s})$	Ratio of Final to Initial Momentum	Momentum Is Conserved?
1				
2				
3				

Assessments

1. Suppose a physicist fires a projectile toward another projectile of the same mass and shape on a track surface. When the projectiles collide, they bounce and move away from each other at the same initial speeds, but in the opposite directions. TEKS (6)(D)
    a. Does this scenario describe an elastic collision? Why or why not?
    b. Does conservation of energy apply to this system? Why or why not?
    c. Does conservation of momentum apply to this system? Why or why not?
2. In an elastic collision, two objects of the same mass moving at the same speed collide and move in opposite directions. The magnitude of the velocity of the first object is .... as the magnitude of the velocity of the second. TEKS (6)(D)
3. In an elastic collision, an object with momentum collides with another object that is at rest. After the collision, the momentum of the first moving object:
    a. Remains unchanged, because it is conserved.
    b. Doubles, because it gains momentum from the second object.
    c. Decreases, because the other object gains some of its momentum in the elastic collision.
    d. Stays the same, because it sticks to the other object and they move together. TEKS (6)(D)

Activity 2: Inelastic Collisions

When two peaches collide, they may stick together, or they may bump and deform, then separate and move apart. We define this interaction as an **inelastic collision** because kinetic energy is not conserved; rather, energy is lost as the peaches crumple together during the collision. The final kinetic energy is less than the sum of the initial kinetic energies. If the peaches stick and move together after the collision, we say this is a perfectly inelastic collision. Not all inelastic collisions occur with objects that stick together. In reality, most collisions are somewhere between elastic and perfectly inelastic collisions.

In this activity, you will measure the velocities of cars undergoing inelastic collisions on a nearly frictionless track. You will calculate momentum before and after the collision to test whether momentum is conserved in the collision. Then you will change the mass and the velocity of the cars to observe how these variables affect conservation of kinetic energy and momentum.

Safety Precautions

- Place the cars gently into the track from above to avoid scratching the surface.
- Do not move the cars horizontally unless the compressed air supply is turned on.
- Wear safety goggles to prevent eye injuries from particles released during collisions.
- Do not throw or propel cars off the track or toward other students.

For this activity you will need the following:

- One frictionless compressed air track with two encoder strips along the length of the track, or a smooth track-like surface (such as a dry-erase board with two parallel meter sticks serving as edges for a track). A commercially available air track is the Vernier Dynamics Cart and Track System with Motion Encoder (Order Code DTS-EC).

  See: http://www.vernier.com/files/manuals/dts-ec.pdf.
- Two dynamic cars with optical encoder and infrared (IR) transmitter (with the same initial mass, and removable weights to attach to cars), and magnets to attach to each car or opposite-polarity Velcro on square bumper for improvised cars.
- One track with two encoder strips along the full length of the track.
- Two IR receivers attached to each end of the track (or four digital timers precise to one tenth or one hundredth of a second).

- <u>Additional items needed in order to improvise a track</u>:
    - One roll of masking tape to hold meter sticks in place on track surface
    - Three digital timers precise to one tenth or one hundredth of a second
    - Two-meter-long smooth boards for incline ramps
    - Two meter sticks marked in cm
    - One cardboard box or stack of books to elevate ramp

For this activity you will work *in groups of four* (two to handle the cars and two to run the data collection from the sensors)

Procedure

Step 1: Set up the air track according to your teacher's directions. For the Vernier equipment, there are two encoder strips, one on each side of the track, with the receiver on the same side as the encoder strip for the car closest to it. The encoder strip consists of alternating black and white bars with a 4 mm period, allowing the optical sensor to detect the passage of the bars as the car moves. No alignments or adjustments are necessary, as one receiver attaches firmly to each end of the track. Consider reversing the direction of one receiver so that the same direction is positive for each system. Put the cars together, and zero both systems. This will put the cars on the same coordinate system; if they move together in contact, their position readings will be the same. Connect the receiver to an interface such as a LabQuest 2. If using a computer, connect the interface to your computer and launch Logger Pro software.

Step 2: Attach magnets of opposite polarity to the ends of the cars that will face each other during the collision. On improvised cars, attach opposite-polarity Velcro strips.

Step 3: Determine the mass of each of the two cars on a balance and record in Table 8.5.

Step 4: Turn on the car by pressing the power button. The car will glow blue when power is on. Place the car down vertically onto the track, wheels in grooves, with the blue light facing the receiver.

Step 5: Launch the cars near the center of the air track. Give each car a push, directing it away from the other car. Start with a slow, gentle push for best results. The cars will bounce off the end of the track and rebound, reaching a stable speed before colliding with each other (Figure 8.2). Be sure to stay out of the way of the motion sensors. Continue taking data as the cars stick to one another and move toward one end of the track. With two sensors appropriately placed on the underside of the cars, a narrow infrared beam transmits motion data from the car sensors to the nearest receiver and can detect a change in position with 1 mm resolution, as well as the direction of travel. Then hit the *Stop* button, using the cursor.

Step 6: The computer will calculate and plot the change in velocity of each car. Record each velocity before the collision as the initial velocity in Table 8.5.

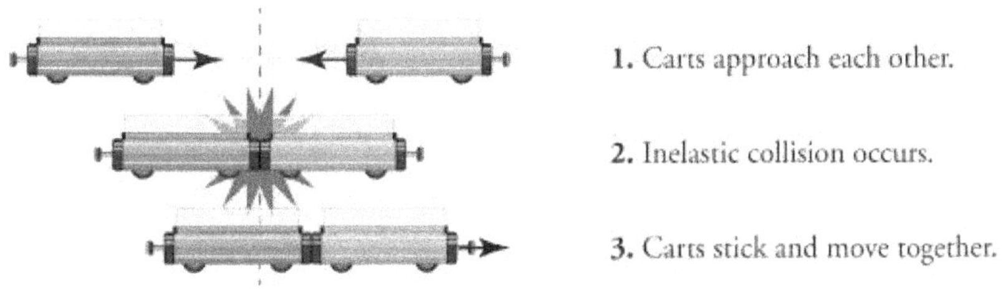

Figure 8.2: *In the three stages of a completely inelastic collision, cars approach each other, collide, stick together (losing kinetic energy in the process), and move together afterward.*

Step 7: The computer will calculate and plot the change in velocity and direction of travel after the collision. The magnets will cause the cars to stick and move together after they collide. Record the velocity and the direction both cars move after they collide as the final velocity in Table 8.5. *Note*—for improvised experiments, students will need to measure times with their digital timers as the cars move away from their predetermined collision points. Students will divide the distance by time to obtain the final velocities.

Step 8: Repeat Steps 3-7, but change the initial velocity of one of the cars by giving the car a harder initial push. The car should move about twice as fast as the initial original run. The exact speed is not important, since the computer will calculate the velocity. Record as Trial 2 in Table 8.5.

Step 9: Repeat Steps 6-8, but change the mass of one of the cars (by adding up to four 125 g weights) and record as Trial 3 in Table 8.5.

Step 10: Calculate initial kinetic energy for each car $(KE = \frac{1}{2}mv^2)$, using the velocities immediately before the collision. Record in Table 8.6.

Step 11: Calculate final kinetic energy $(KE = \frac{1}{2}mv^2)$, using $(m = m_1 + m_2)$ for the total mass and using the final velocity for the two cars joined together immediately after the collision event. Record in Table 8.6 for each respective trial run.

Step 12: Calculate initial and final momentum $(p = mv)$ using $(m = m_1 + m_2)$ for the combined mass, and using the final velocity for the two cars joined together immediately after the collision event. Record in Table 8.7.

Step 13: Record the total kinetic energy before and after colliding in Table 8.8 using the equations:

$$\text{Initial total KE} = \tfrac{1}{2}m_1v_{1i}^2 + \tfrac{1}{2}m_2v_{2i}^2$$
$$\text{Final total KE} = \tfrac{1}{2}(m_1 + m_2)v_f^2$$

Step 14: Record the total momentum before and after colliding in Table 8.9 using the equations:

$$\text{Initial momentum} = m_1v_{1i} + m_2v_{2i}$$
$$\text{Final momentum} = (m_1 + m_2)v_f$$

Step 15: Compare the total kinetic energy before and after the inelastic collisions by recording in Table 8.8 the ratio of final to initial total kinetic energy.

Step 16: Compare the total momentum before and after the inelastic collisions by recording in

Table 8.9 the ratio of final to initial total momentum.

Step 17: Compare Trial 1 to Trial 2 in all experiments to determine whether changing the velocity of a car changes the conservation of energy, and record the results in Table 8.8. Also determine whether changing the velocity changes the conservation of momentum, and record the results in Table 8.9.

Step 18: Compare Trial 1 to Trial 3 in all experiments to determine whether changing the mass of a car changes the conservation of energy, and record the results in Table 8.8. Also determine whether changing the velocity changes the conservation of momentum, and record the results in Table 8.9.

Data and Observations

Table 8.5 Inelastic Collision: Mass and Velocity Measurements

Trial Run and Car Number	Mass of Car (kg)	Initial Velocity (m/s)	Final Velocity (m/s) and Direction of Both Cars Afterwards
1 - Car A			
1 - Car B			
2 - Car A			
2 - Car B			
3 - Car A			
3 - Car B plus 500 g			

Table 8.6 Inelastic Collision: Kinetic Energy

Trial Run and Car Number	Initial Kinetic Energy (J)	Final Kinetic Energy (J)	Direction of Motion
1 - Car A			
1 - Car B			
2 - Car A			
2 - Car B			
3 - Car A			
3 - Car B plus 500 g			

Table 8.7 Inelastic Collision: Momentum

Trial Run and Car Number	Initial Momentum (kg ms-1)	Final Momentum (kg mec-1)	Direction of Motion
1 - Car A			
1 - Car B			
2 - Car A			
2 - Car B			
3 - Car A			
3 - Car B plus 500 g			

Table 8.8 Conservation of Kinetic Energy in Inelastic Collisions

Trial Number	Total Initial Kinetic Energy (J)	Total Final Kinetic Energy (J)	Ratio of Final to Initial Total Kinetic Energy	Kinetic Energy is Conserved?
1				
2				
3				

Table 8.9 Conservation of Momentum in Inelastic Collisions

Trial Number	Total Initial Momentum $(kg \cdot m/sec)$	Total Final Momentum $(kg \cdot m/sec)$	Ratio of Final to Initial Total Momentum	Momentum is Conserved?
1				
2				
3				

Assessments

1. A physicist sets up a test where he fires a projectile with a mass of 0.25 kg toward a board with a mass of 8.5 kg. The projectile enters the board and the board and projectile move together in the direction the projectile was initially traveling. TEKS (6)(D)
    a. Is this an elastic or an inelastic collision? Why?
    b. Does conservation of energy apply to this system? Why or why not?
    c. Does conservation of momentum apply to this system? Why or why not?
2. In an inelastic collision, an object moving right collides with an object of the same mass moving left and they stick together after the collision. If both objects move left after the collision, we could say that the first object was traveling ....than the second. TEKS (6)(D)
3. In an inelastic collision, an object with momentum collides with another object that is moving at a slower velocity but in the same direction as the first object. After the collision, the momentum of the first moving object:
    a. Increases unchanged, because it is conserved.
    b. Reverses, because it moves backward after colliding with the second object.
    c. Decreases, because the other object gains loses of its momentum and reverses.
    d. Is the same as that of the second object, because the two objects stick and move together. TEKS (6)(D)

Lab 9: Thermodynamics

Many people find the difference between temperature and heat confusing. **Temperature** is a measurement of thermal energy. Thermal energy is a measure of the average kinetic energy of particles in a substance. The particles have kinetic energy because they are moving around; the faster they move, the more kinetic energy (and thermal energy) they have. Even the atoms in a solid vibrate back and forth and so have some motion. **Heat** is the transfer of thermal energy from one substance to another. For example, suppose a glass of water has a temperature of $23°C$. When the glass is placed in a freezer, heat flows from the water to the cooler surroundings. Measuring the temperature of the water at different times allows us to determine how much heat has been transferred.

In this lab, you will learn to measure temperatures accurately, and then use that skill to determine how much heat is transferred when objects at different temperatures interact.

<u>In this lab you will learn:</u>

- how to measure temperature properly by using a thermometer for scientific data collection;
- how to convert temperatures between Celsius, Fahrenheit and Kelvin scales;
- how to calculate the amount of heat transferred based on temperature change.

Activity 1: Measuring Temperature TEKS (6)(G) TEKS (6)(E)

All temperatures are measured relative to some baseline. In the United States, three temperature scales are commonly used. In 1714, Daniel Fahrenheit invented the first reliable thermometer. He chose the temperature $0°F$ to be the coldest temperature because he could easily reproduce this temperature in his lab by mixing salts and ice. Later, the astronomer Anders Celsius invented a different temperature scale based on water; he set $0°C$ as the temperature at which water freezes and $100°C$ as the temperature at which water boils. In both the Celsius and Fahrenheit scales, temperatures are measured in degrees. Measuring a negative temperature simply means that the system being measured has less thermal energy than the chosen zero point on that scale.

Recall that temperature is a measure of thermal energy. At **absolute zero**, a system has no thermal energy, meaning all particles are completely still. (In reality, we cannot achieve absolute zero because we cannot remove all thermal energy.) The Kelvin temperature scale uses absolute zero as its zero point. It is not possible to transfer thermal energy from a system that has none, so absolute zero is the coldest temperature possible, and all Kelvin temperatures are positive. The unit for measuring temperature on this scale is the kelvin (not degrees kelvin, so degree signs are not used with K) and one kelvin is the same magnitude as one degree Celsius. Thermometers can display any of these scales, as shown in Figure 9.1.

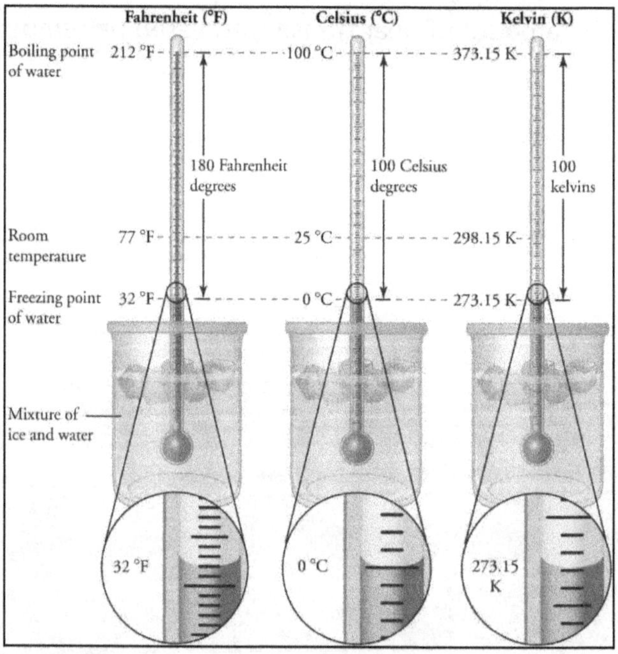

Figure 9.1: *At the freezing point of water, the temperature is* $32\ °F$, $0\ °C$, *or 273.15 K, depending on the scale used.*

Convert between Fahrenheit and Celsius, with the formula

$$T_C = \frac{5}{9}(T_F - 32)\ .$$

Convert between Celsius and Kelvin by using

$$T_K = T_C + 273.15\,.$$

Safety Precautions

- Inform your teacher immediately of any broken glassware, to avoid injuries.
- Clean up any spilled water or other fluids to prevent other people from slipping.
- If your thermometer contains silver-colored liquid, it is a mercury thermometer. Be extra careful with it. If the thermometer breaks, immediately inform a teacher, because mercury requires specific clean-up protocols.

For this activity you will need the following:

- Water
- Balance
- Beaker or graduated cylinder
- Thermometer
- Ice
- Three cups
- Timer

For this activity you will work *in pairs*.

Procedure

Step 1: Use the balance to measure approximately 50 g of ice into one cup and 10 g of ice into a second cup. Use the beaker to measure 50 mL of water (at room temperature) into the first cup, 90 mL of water into the second cup, and 100 mL of water into the third (empty) cup.

Step 2: Make sure there is a continuous column of liquid (no breaks or bubbles) inside your thermometer. Place the thermometer in the first cup. Swirl gently to make sure the ice water is thoroughly mixed. When the temperature stops changing, record the value in Table 9.1 below. Make sure you are counting the tick marks on your thermometer correctly (Figure 9.2).

Step 3: Convert whichever temperature scale your thermometer reads to both of the other scales, using the equations in the introduction.

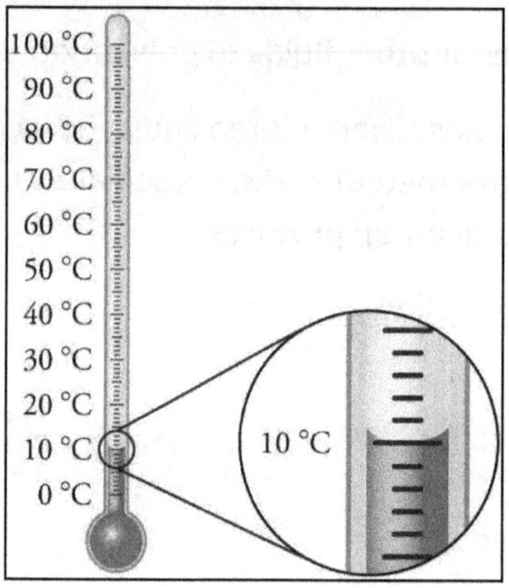

Figure 9.2: *On this particular thermometer, there are five small tick marks between each labeled in* $10°$ *increment, so each small line represents 2°. The red liquid (alcohol) is at the* $10°C$ *line.*

Data and Observations

<u>Table 9.1 Temperature of Ice and Water</u>

	Temp. in °C	Temp. in °F	Temp. in K
Cup 1: ~50 g ice			
Cup 2: ~10 g ice			
Cup 3: no ice			

Assessments

1. Explain the relationship between molecular movement and the temperature in each cup. Where are the particles moving slowest? How do you know? TEKS (6)(E)
2. Select the best answer from the underlined choices.
    a. The ice was <u>hotter than/cooler than/the same temperature as</u> the water before mixing.
    b. In cup 1, the heat <u>flowed from the ice to the water/flowed from the water to the ice/did not change.</u>
    c. The water temperature <u>rose/fell/stayed the same</u> and the ice temperature <u>rose/fell/stayed the same.</u>
    d. The total energy of the water and ice system <u>was conserved/increased/decreased.</u> TEKS (6)(G)
3. A website reports that the temperature on Pluto is very, very cold (–230 K). Explain whether this seems correct or whether it might be a mistake. a typo. TEKS (6)(G)
4. Predict the estimated temperature, in degrees Celsius, if 25 g of ice are mixed with 75 mL of room temperature ($25°C$) water. TEKS (6)(G)

Activity 2: Calorimetry TEKS (6)(G)

Heat always transfers energy from higher temperature areas to lower temperature areas. Consistent with the second law of thermodynamics, heat can only flow in the direction that increases the distribution of thermal energy, increasing entropy. Objects cooler than their surroundings absorb heat, and as the temperature of an object increases, the temperature of the object's surroundings decreases. Eventually the object and its surroundings are at the same temperature, a state known as **thermal equilibrium**. At thermal equilibrium, no exchange of heat is observed. Calorimetry is the study of the flow of heat and is used in many real-world applications, from determining how many calories are in a cookie to studying the quality of concrete.

In this activity, you will investigate the fundamentals of calorimetry. The amount of heat transferred (*Q*) is an amount of energy, measured in joules. You can calculate the amount of heat transferred by measuring the change in temperature of a system in degrees Celsius. The equation that relates heat and temperature change (ΔT) is

$$Q = mc\Delta T \ .$$

In this equation, *m* is the mass of the substance (in grams) you are measuring and *c* is the specific heat of that substance. When measuring the temperature of water, the density

$(\rho = 1 \text{ g/mL})$ and measured volume are typically used to calculate the mass of water. You can use the equation for density and rearrange it to solve for mass in the equation below.

$$m = \rho \times V$$

Specific heat (*c*) is the amount of energy needed to raise the temperature of 1.00 kg of a sample by $1.00\ °\text{C}$. Because you are measuring the temperature of water, you should use the value for water, $c = 4184 \text{ J/(kg}\cdot°\text{C)}$

<u>Safety Precautions</u>

- Inform your teacher immediately of any broken glassware, to avoid injuries.
- Clean up any spilled water or other fluids to prevent other people from slipping.
- If your thermometer contains silver-colored liquid, it is a mercury thermometer. Be extra careful with it. If the thermometer breaks, immediately inform a teacher, because mercury requires specific clean-up protocols.
- Hot water is hot! Avoid burns by being cautious when removing objects from hot water and use clamps to avoid spills.

<u>For this activity you will need the following:</u>

- ~25 g of metal (10 pennies or a lead sinker from a fishing store work well)
- Large test tube
- Test tube holder
- Water
- Hot plate or Bunsen burner with stand
- Two foam cups
- Graduated cylinder
- Two beakers
- Thermometer
- Clamp for beaker
- Thermometer clamp and stand
- Piece of cardboard or lid for foam cup

For this activity, you will work *in pairs*.

Procedure

Step 1: Fill a beaker approximately 2/3 full of water and position it over a hot plate or Bunsen burner. Clamp the beaker in place. Place the metal object in a clean, dry test tube and use the test tube holder to position this tube in the beaker of water. Heat the water until it is gently boiling. Continue heating for 10 minutes to ensure that the metal comes to thermal equilibrium with the hot water. Do not let the beaker boil to dryness; add more water and reduce heat if the water level goes below the level of the metal in the test tube.

Step 2: While the metal is heating, assemble a coffee cup calorimeter as shown in Figure 9.3. Using a beaker to provide stability, stack both foam cups together to provide good insulation. Measure 40 mL of water and transfer it to the calorimeter (the water must completely cover your piece of metal). You may need to use a different amount of water, depending on your cup and metal size. Be sure to record the precise volume of water used.

Step 3: Position a clamp for the thermometer so that it is submerged in the water. Use a lid or a small piece of cardboard to cover most of the cup opening. Record the initial water temperature, T_i, in the chart.

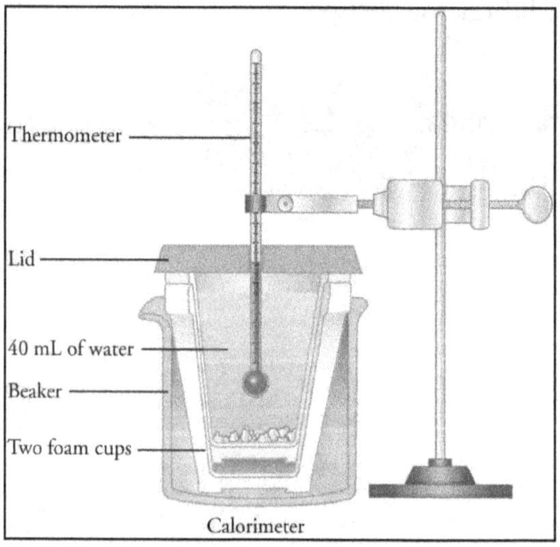

***Figure 9.3:** In the coffee cup calorimeter, heat is transferred from the metal to the water. Note that the thermometer is clamped so that it is not touching the metal.*

Step 4: Temporarily remove the lid from the calorimeter. Take the test tube out of the boiling water using the test tube holder and quickly and carefully pour the metal from the test tube into the calorimeter. Adjust the position of the thermometer so that it is above the metal but still submerged in the water. Replace the calorimeter lid and gently swirl the cup, being careful not to break the thermometer. Record the highest temperature that the thermometer reaches as T_f.

Step 5: Fill in Table 9.2 by calculating the mass of the water in the calorimeter, using the water's density. Then use the equation and value provided in the introduction to calculate the heat transferred to the water.

Data and Observations

Table 9.2 Transfer of Heat

T_i of water in °C	
T_f of water in °C	
ΔT	
Volume of water in mL	
Mass of water in g	
Mass of water in kg	
Heat transferred, Q	

Assessments

1. In this experiment, the hot metal transfers heat to the cooler water until the system comes to thermal equilibrium. However, most experimenters observe that the temperature in the calorimeter increases and then decreases, instead of remaining constant. Explain this observation. TEKS (6)(G)
2. Metals typically have significantly lower specific heat values than water. Recall that specific heat is the amount of energy required to increase the kinetic energy (measured as temperature) by a certain amount. Therefore, if you add 1000 J of heat to 500 g each of water and metal, the metal will increase in temperature significantly more than the water. Thinking at a molecular level, explain this trend. Then explain why it is easier to change the temperature of a solid piece of metal compared with liquid water. TEKS (6)(E)
3. Which has a larger temperature change: the water in the calorimeter or the metal? (*Hint*: the metal's starting temperature was approximately the temperature of boiling water, 100°C). Which has a larger amount of heat transferred? Which has a larger specific heat? Explain each of your answers. TEKS (6)(G)

Lab 10: Waves

From ocean waves to sound waves, waves constantly surround us. Waves can be described using **frequency** (the number of waves passing per time period), **wavelength** (the length of the wave from crest to crest or trough to trough) and **amplitude** (the height of the wave). Waves interact with matter through **reflection,** as they bounce off surfaces, and **refraction,** as they bend through different **media**. There are two main types of waves: **transverse** waves, where the particles move perpendicular to the wave, and **longitudinal** waves where the particles move parallel to the motion of the wave.

In this experiment you will use spring toys to create a transverse wave and a longitudinal wave. Using the spring toy, you will see how waves reflect back when encountering a fixed end and loose end. You will also use a modified ripple tank to observe how water waves interact with boundaries. In this lab you will be making observations of the waves.

In this lab you will learn:

- how a spring toy can create a longitudinal wave and a transverse wave;
- how waves reflect off loose and fixed ends of a spring toy;
- how waves interfere with one another.

Activity 1: Producing Waves TEKS (7)(C)

Waves are classified into two different types: transverse and longitudinal. In this activity you will create **pulses**—one wavelength—of these types of waves with a spring toy. In a transverse wave, the particles move perpendicular to the direction of motion of the wave. For a longitudinal wave, the particles move parallel to the direction of the wave. As you move the spring toy, the coils act as the particles and move perpendicular or parallel to the wave created.

Safety Precautions

- Do not release stretched spring toys to avoid being hit by a rebounding spring
- Gather the spring toys back together gently to avoid tangling

For this activity you will need the following:

- Long spring toy

For this activity you will work *in pairs*.

Procedure

Step 1: Have one student at each end and stretch the spring toy between themselves so that the coils are about an inch apart.

Step 2: Have one partner move the spring toy from side to side once (create one crest) to produce a single transverse wave pulse. Record your observations in Table 10.1.

Step 3: Have one partner flick along the length of the spring toy once to create a single longitudinal wave pulse (Figure 10.1). Record your observations in Table 10.1.

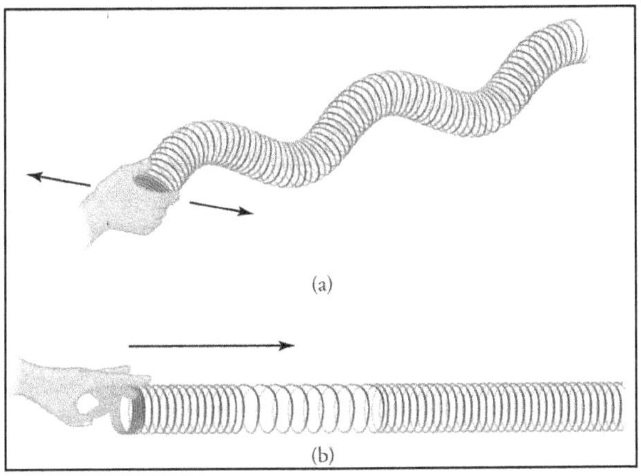

Figure 10.1: *Quickly flick the coils in a direction parallel to the spring toy.*

Data and Observations

Table 10.1: Wave Types

Wave Pulse Type	Diagram of Motion (Be sure to draw an arrow to show the direction of motion)	Description of Motion
Transverse pulse		
Longitudinal pulse		

Assessments

1. A particle in a longitudinal wave moves .... to the wave, whereas particles in a transverse wave move ...... to the motion of the wave. TEKS (7)(C)
2. A boat is floating on the water. As a wave passes by the boat moves upward and then downward. What type of wave of this? Explain how you know. TEKS (7)(C)
3. As a tuning fork vibrates, it causes air particles around them to move back and forth. These particles collide with each other, sending the wave toward your ear. What type of wave does the sound of a tuning fork produce? TEKS (7)(C)

Activity 2: Wave Reflection TEKS (7)(D)

Waves constantly interact with different surfaces. As waves strike surfaces, they behave differently depending on the type of surface they encounter. If the wave encounters a medium or surface that is denser than the medium the wave was traveling through, the wave will reflect back inverted, upside down. If the wave encounters a medium or surface that is less dense, the wave reflects back the same way, upright. In this activity you will simulate a denser surface, or fixed end, with a textbook and a less dense surface or loose end, with a string.

Safety Precautions

- Do not release stretched spring toys to avoid being hit by a rebounding spring
- Gather the spring toy back together gently to avoid tangling

For this activity you will need the following:

- Long spring toy
- String, 50 cm long
- Textbook

For this activity you will work *in pairs*.

Procedure

Step 1: With a student at each end, stretch the spring toy between themselves so that the coils are about an inch apart.

Step 2: Hold the last coil of the spring in the cover of a textbook so that the front cover faces the spring (Figure 10.2). You may need to close the book in order to get a firm hold to create a fixed end for the wave to reflect off.

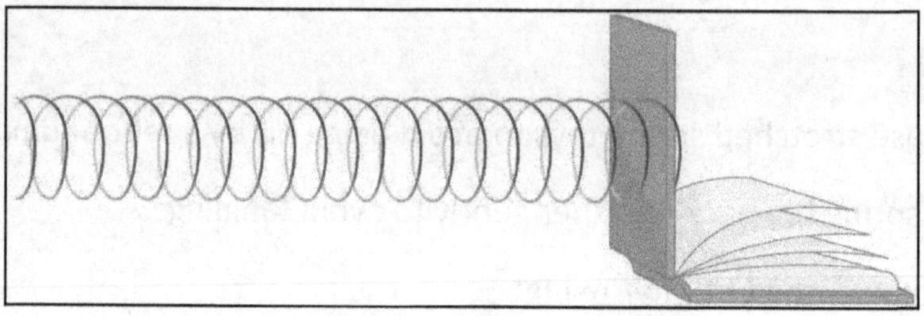

Figure 10.2: *Insert the last coil inside the cover of a textbook to create a fixed end.*

Step 3: Have one partner move the spring toy quickly to the right and back to the original position once to produce a wave pulse.

Step 4: Observe how the wave pulse returns after it interacts with the book. Does it return on the same side or the opposite side? Record your observations in Table 10.2. Be sure to draw an arrow to show the direction of motion.

Step 5: Attach a string, about 12 inches long, to the last coil of the spring toy and hold the end of the string. This creates a loose end for the wave to reflect off.

Step 6: Have one partner move the spring toy quickly to the right and back to the original position once to produce a wave pulse.

Step 7: Observe how the wave pulse returns after it reaches the end of the spring. Does it return on the same side or the opposite side? Record your observations in Table 10.2. Be sure to draw an arrow to show the direction of motion.

Data and Observations

Table 10.2: Reflection of Waves

Reflective Surface	Sketch of Wave Before	Sketch of Reflected Wave	Description of Motion
Fixed end (textbook)			
Loose end (string)			

Assessments

1. A wave pulse is reflected back upright. Was this barrier a loose end or a fixed end? How do you know? TEKS (7)(D)
2. A wave hits a boundary that is a denser medium. Is the reflected wave upright or inverted? Use your observations to explain your prediction. TEKS (7)(D)
3. A wave pulse with an amplitude of 3 cm is traveling to the right and is reflected off a fixed end. Draw the graph of the reflected wave. Be sure to draw an arrow indicating the direction of travel. TEKS (7)(D)

Activity 3: Waves in Motion TEKS (7)(D)

Waves often overlap and interfere with one another. This interference can be constructive or destructive. **Constructive interference** happens when two crests or two troughs align and result in a larger wave amplitude. **Destructive interference** happens when a crest and a trough align and result in a smaller amplitude. Ripple tanks are a great way to visualize how waves interfere. In a traditional ripple tank, a light is shown over a tank of water. As waves are generated in the tank, shadows indicate waves with higher amplitude while lighter spots indicate waves with less amplitude. In our tray we will see high amplitudes as darker colors and lower amplitudes with lighter color.

Safety Precautions:

- Keep the water tray away from electrical sockets to prevent electric shock
- Clean up any spilled water to prevent other people from slipping and falling

For this activity you will need the following:

- Transparent tray with colored water
- Graph paper
- Note card
- Blocks

For this activity you will work *in pairs*.

Procedure

Step 1: Fill the tray with about half a centimeter of colored water. The colored water will help you visualize the crests and troughs of the waves. Put graph paper under the bottom of the tray so you can see the grid. This will help you judge the size and speed of the waves.

Step 2: Place a note card or other straight object on one side of the tray. Then move it back and forth continuously at a constant rate to produce waves that hit the opposite wall. Sketch the waves in Table 10.3 as they travel across the tray. Be sure to draw an arrow to show the direction of motion. Your sketch should look something like Figure 10.3.

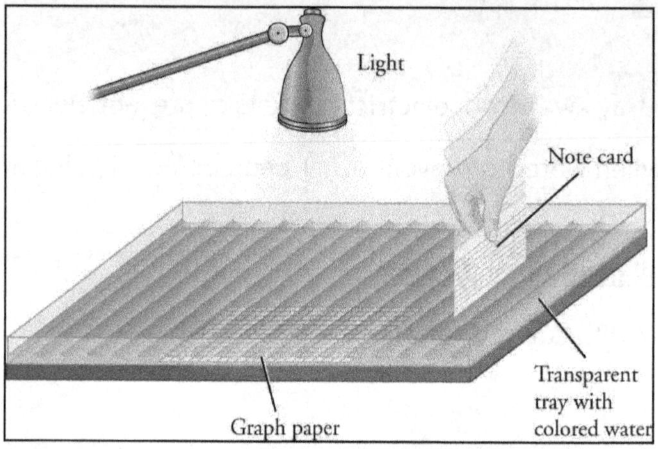

Figure 10.3: *Place the notecard into the tray and move it back and forth to create waves. You should create dark and light areas inside the tray that indicate the crests and troughs of the waves.*

Step 3: Add blocks of wood toward the middle of the tray. Place them so that there is about a one-centimeter slit between them. Use the notecard to create waves and observe how the waves travel through the slit. What happens to the waves after they travel through the slit? Sketch the waves in Table 10.3 as they travel across the tray and through the slit 3. Be sure to draw an arrow to show the direction of motion.

Step 4: Remove the blocks. Place the index card in the tray at a 45° angle and move it back and forth in the tray. Sketch the waves in Table 10.3 as they travel across the tray. Be sure to draw an arrow to show the direction of motion.

Data and Observations

Table 10.3: Waves in a Tank

Angle of Card	Sketch of Waves in Tray	Description of Motion
Parallel to wall		
Through a slit		
At 45° angle to wall		

Assessments

1. As the waves move across the tray and reflect off the wall, do they keep the same speed or change speed? If the speed decreased, what would happen to the wavelength if the frequency remained the same? TEKS (7)(D)
2. As the waves reflect off the wall, describe what happens to the amplitude of the wave when

    a. The crest of the reflected wave crosses the crest of the oncoming wave.

    b. The crest of the reflected wave crosses the trough of the oncoming wave.

    TEKS (7)(D)
3. The waves created in the tank are longitudinal waves, like sound waves. Use your observations to explain why you can hear sound around a corner. TEKS (7)(D)

Lab 11: Sound Waves

Sound waves are a mechanical method of energy transfer. More specifically, a sound wave is the pattern of vibration caused by energy moving through a form of matter referred to as a *medium*. Sound is transmitted through the matter away from the source of the sound. When we hear, we are hearing these vibrations travelling through the air.

Sound vibrations travel through matter in waves. Therefore, sound waves can be described by using **frequency** (the number of waves passing per time period), **wavelength** (the length of the wave from crest to crest or trough to trough), and **amplitude** (the height of the wave). When two sound waves add up so their crests meet, it is called **constructive interference** and the waves increase in amplitude. **Destructive interference** occurs when the crest of one wave meets the trough of another; in this case, the total amplitude of the wave decreases. When sound vibrates without any interference from external forces, it vibrates at its **natural frequency**. **Resonance** occurs when one object that is vibrating at its natural frequency causes another object to also vibrate at its natural frequency. **Beats** are produced when two waves with a small difference in frequency but identical amplitudes are superimposed. The waves alternate between constructive and destructive interference.

In this lab you will learn:

- how to determine the resonating frequency of various tuning forks;
- how to create sound in a bottle or glass to replicate a song with simple notes;
- how to use a tuning fork to determine the speed of sound in water.

Activity 1: Resonance TEKS (7)(D)

Resonance is forced oscillatory motion that lines up with the natural oscillation frequency of a system. Reinforcing the motion with force in the correct direction and frequency produces an increase in amplitude. Resonance occurs when the natural vibration rates of two objects are the same or when the natural vibration rate of one is sympathetic to the other. The resonant frequency is a natural frequency of vibration determined by the physical characteristics of the vibrating object. Forced vibration at this natural frequency causes resonance.

Tuning forks consist of a handle and two prongs, or tines. The tines begin to vibrate when they are struck, and the back-and-forth vibration disturbs surrounding air molecules (Figure 11.1). The tine stretches outward and compresses surrounding air molecules to create a high-pressure region (compression) next to the time. The time then moves inward and the surrounding air expands to produce a low-pressure region (rarefaction) next to the time. These compressions and rarefactions are transported to the air around the fork and constitute the sound signal. The air molecules vibrate back and forth continuously from their original position and have no net displacement. They are disturbed only temporarily. In this activity, you will force tuning forks to vibrate and produce resonance.

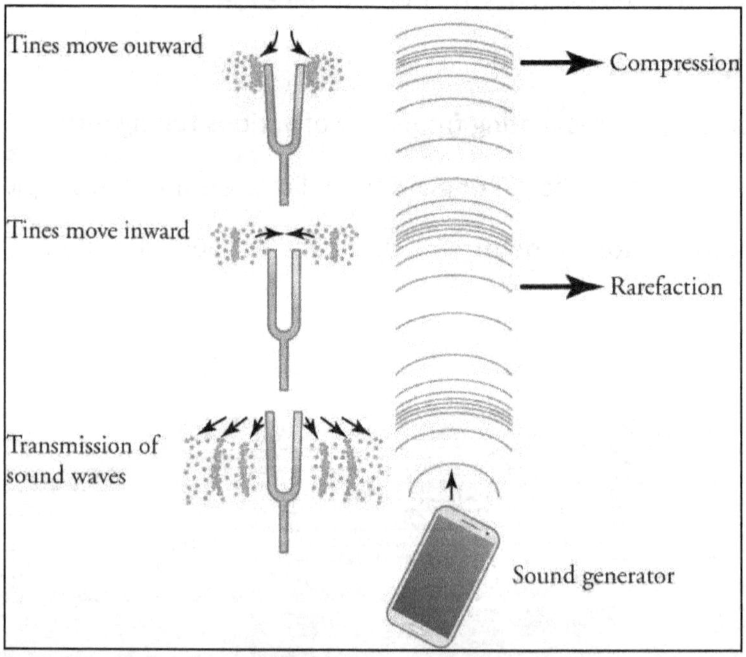

Figure 11.1: *Sound from the sound generator in the phone produces waves that move the tines of the tuning fork.*

Safety Precautions

- Do not try to carry too many materials at once. Make multiple trips if needed.
- Do not drop the tuning fork; the times are easily bent.
- Be careful with the glass bottles; do not drop them.
- Do not turn up the sound generator too high.

For this activity you will need the following:

- Two tuning forks of different wavelength
- Sound generator
- Two identical glass bottles
- Golf ball

For this activity you will work *in pairs*.

Procedure

Step 1: Fill one bottle with water about halfway. Strike a tuning fork on the golf ball and hold it over the bottle. If you do not hear anything, add or remove water until you hear a resonant sound after the tuning fork is struck. When the tuning fork causes the air in the water bottle to vibrate, you are hearing resonance. Record in Table 11.1 the frequency and length of the tuning fork, and the height of the water.

Step 2: Fill the second bottle with water about halfway. Strike the second tuning fork on the golf ball and hold it over the bottle. If you do not hear anything, add or remove water until you hear a resonant sound after the tuning fork is struck. Record in Table 11.1 the frequency and length of the tuning fork and the height of the water.

Step 3: Select a bottle and tuning fork that produce resonance. Note the frequency of the fork. Open the sound generator, select *sine*, and enter the frequency of the tuning fork.

Step 4: Tap the tuning fork and hold it over the bottle. Listen to the tones from the fork and the sound generator. If the frequencies are matched accurately, you will hear a boost in sound. Adjust the sound generator until you hear a boost in sound. Note that in Table 11.2 any variance in the frequency of the tuning fork and the frequency of the sound generator.

Step 5: Alter the frequency on the sound generator by 1–4 Hertz. Note that in Table 11.2 whether you hear any beats or resonance.

Step 6: Set the frequency on the sound generator to be half or double the frequency of the tuning fork. Play the sound generator and tuning fork in tandem. Note any observations in Table 11.2.

Step 7: Set the frequency on the sound generator to be half the frequency of the tuning fork. What do you hear? Note your observations in Table 11.2.

Data and Observations

Table 11.1: Frequency in Bottles of Water

	Frequency	Tuning Fork Length	Height of Water
Fork 1			
Fork 2			

Table 11.2: Observations in the Bottle

Step	Observations
Tuning fork + sound generator	
Beats	
Half or double frequency	
Half frequency	

Assessments

1. Consider the difference between the sound in the bottle and the sound at your ear. Do you notice any difference with distance away from the source? What does the sound do? TEKS (7)(D)
2. In Step 6, you were encouraged to halve or double the selected frequency and play the sound generator and tuning fork in tandem. What did you hear? What wave interactions likely caused this to happen? TEKS (7)(D)
3. What are the best groupings and distances apart to place sound sources so resonance is achieved? TEKS (7)(D)

Activity 2: Speed of Sound TEKS (7)(B)

Imagine sound waves traveling through a tube that is closed at one end. The sound waves reflect off the closed end and travel back toward the source. As they reach the incoming waves traveling in the opposite direction, the two sets of waves interfere with each other constructively. A node occurs at the closed end of the tube and an antinode is created at the open end, depending on the wavelength (Figure 11.2). A node is the point along a standing wave where the wave has minimum amplitude; the antinode is where the amplitude is at its maximum. Wave speed depends on wavelength and frequency

$$v = \lambda f,$$

where v is wave speed in meters per second (m/s), λ is wavelength in meters (m), and f is frequency in hertz (Hz). A hertz is the measure of an event that repeats once per second. You will use this formula to determine the speed of sound in a tube.

Vibrations inside a tube form a **standing wave** when the wave reflects off the end of the tube. This standing wave is what makes notes we can hear. Blowing on an instrument causes the air molecules to vibrate, which creates the waves that fit in the tube and cause resonance. The **fundamental wave** is the longest wave that fits in the tube. Frequencies that are greater than the fundamental frequency are called **overtones**; **harmonics** are multiples of the fundamental wave. Let us assume the walls of the tube are rigid and do not flex under the pressure variations of the wave. Also, let us assume the walls of the tube are smooth and do not cause attenuation, or a gradual loss of intensity. Intensity is a measure of the sound's energy, but this topic is not addressed in this lab.

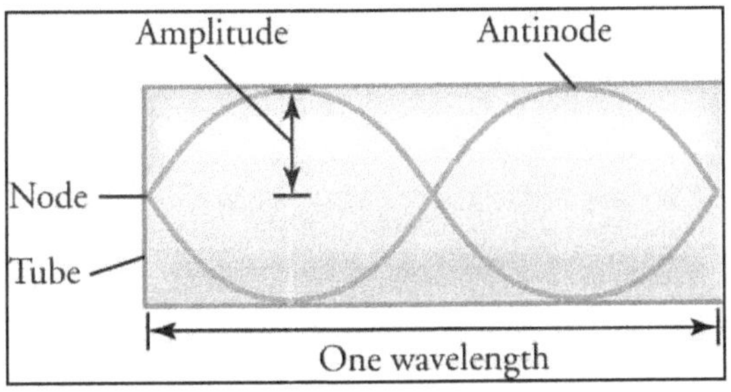

***Figure 11.2**: A standing wave is generated in a tube with one closed end.*

<u>Safety Precautions</u>

- Place objects gently into the resonance tube to avoid spilling.
- To avoid injury, inform your teacher immediately of any broken glassware or water on the floor.
- Clean up any spilled water or other nontoxic fluids to prevent other people from slipping.

For this activity you will need the following:

- Tuning forks of different frequencies or a sound generator (Note—Many phone applications do this.)
- Golf balls if using tuning forks
- Resonance tube filled with water
- Felt-tip marker
- Water
- Meter stick

For this activity you will work *in pairs*.

Procedure

Step 1: Select a tuning fork and strike it on the golf ball, and then hold it about an inch above the open end of the resonance tube.

Step 2: Verify the water level is at the top of the resonance tube, then lower the water level slowly in the tube, as demonstrated by your teacher. Lower the water level until you hear the tube resonate. Use the felt-tip marker to mark the water level on the tube at the point where it gives a resonant vibration. Continue lowering the water level until you have heard and marked as many resonant positions as possible. Note that long wavelengths provide only one or two resonant positions.

Step 3: Record the water level positions for each resonance in Figure 11.3. Sketch the standing wave carefully, as if it were visible. Your sketch should include the actual distances from the top of the tube to each resonant position.

Step 4: Repeat Steps 1 through 3 for two additional tuning forks of different frequencies.

Step 5: Calculate the speed of sound for each tone using the equation $v = \lambda f$ and record it in Table 11.3.

Data and Observations

Table 11.3 Resonant Frequencies

Object	Resonance Frequency (Hz)	Wavelength Determined from Sketch (m)	Wave Speed (m/s)
Sound 1			
Sound 2			
Sound 3			

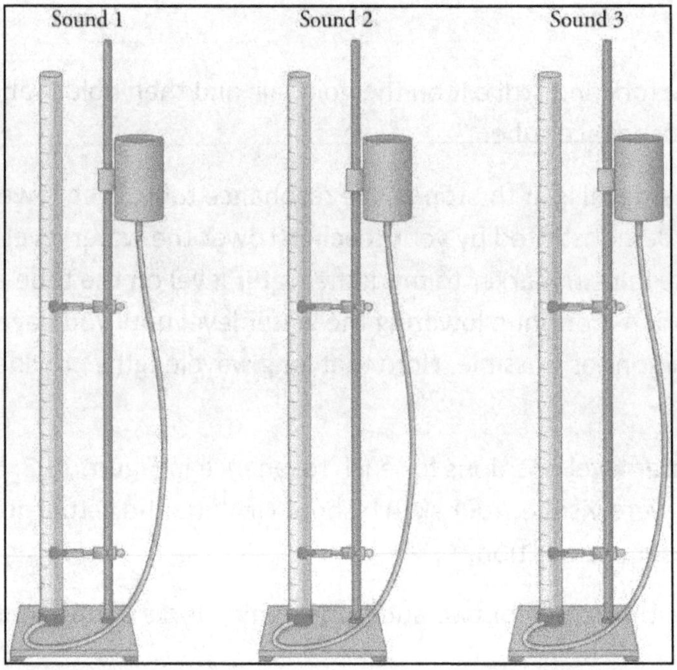

Figure 11.3: *Mark the water-level positions for each observed resonance. Sketch the standing wave as if it were visible and include the actual distances from the top of the tube to each resonant position.*

Assessments

1. The *period* is the time for a particle in the medium to complete one vibrational cycle. Calculate the period of each tuning fork you used in this experiment. Why is the period of an object valuable? TEKS (7)(B)
2. Assume the temperature in your classroom is 20°C and the air is dry. Look up the speed of sound under these conditions in a reference book or on a reputable Internet website. What is the speed of sound under these conditions? Compute the percent difference with each experimental result. TEKS (7)(B)
3. The amplitude of the wave is a measure of its displacement over time. The maximum height observed is the amplitude of a wave. Using the formula $A = \lambda / f$, calculate the amplitude of each tuning fork you used in the experiment. TEKS (7)(B)

Lab 12: Light and Color

Light is an **electromagnetic wave**, which is a wave produced by oscillating electric and magnetic fields. An electromagnetic wave is a particular kind of wave called a **transverse wave**. In a transverse wave, oscillations are perpendicular to the direction of the wave's velocity.

Like all waves, electromagnetic waves are characterized by their speed, **frequency**, and **wavelength**. A wave's frequency measures how often it completes one oscillation in units of Hertz. Wavelength measures the length of one oscillation, in units of meters. All electromagnetic waves travel at the same speed, $c = 3.0 \times 10^8$ m/s, and their velocity, ***c***, is related to their frequency and wavelength by the formula

$$c = \lambda f \, .$$

The energy of an electromagnetic wave is also proportional to its frequency, which is given by

$$E = hf \, ,$$

where $h = 6.6 \times 10^{-34}\ \text{m}^2 \cdot \text{kg/s}$ is Planck's constant.

We can see objects around us, because our eyes detect light. If an object either emits or reflects visible light, we can see it. However, not all electromagnetic waves are visible to humans. Our eyes can detect only a small range of frequencies, and these frequencies make up the visible part of the **electromagnetic spectrum**, shown in Figure 12.1. We cannot see any frequencies below or above the visible range. The particular frequency and wavelength of the light that you are seeing determines its color.

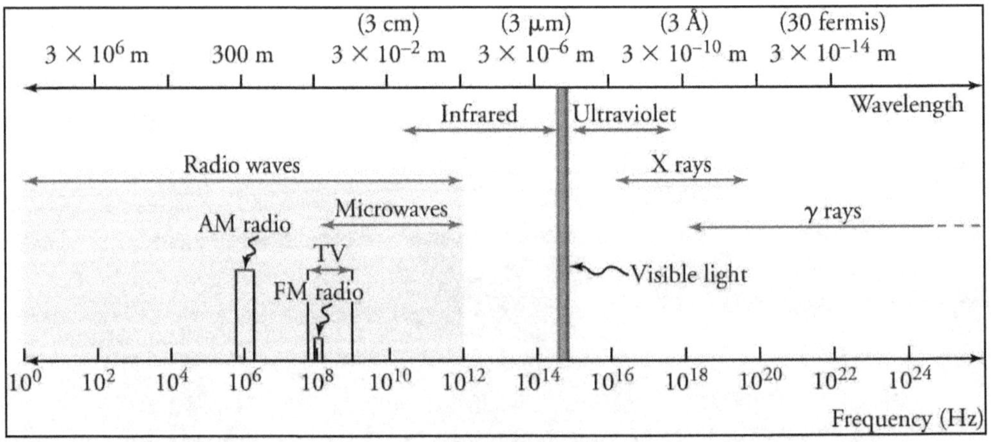

Figure 12.1: *Electromagnetic waves are classified based on their wavelengths. Visible light that we are able to see makes up only a small part of the electromagnetic spectrum.*

In this lab you will learn:

- how to read the emission spectrum of a light source;
- how to identify an element based on its emission spectrum;
- how color mixing of light affects what we see.

Activity 1: Light from an Emission Spectrum TEKS (7)(C), (8)(B)

When you apply a voltage across a tube of gas, the tube gives off light. The electricity excites the electrons in the gas atoms. When the electrons relax back to their original states, they release the extra energy—which is equal to the difference in the energies of the states—as light. The light released due to the transition from a higher to a lower energy state makes up the **emission spectrum** for that particular gas.

Electrons in atoms must be at discrete energy levels. Every element has a unique atomic structure, and the spacing of these electron energy levels is different for every element. Therefore, each different element emits light at a particular set of energies. This leads to the formation of a **discrete spectrum**, which is a spectrum made up of distinct energies, each specific to the element that produced it. In other words, a discrete spectrum can be used to *fingerprint* the presence of certain elements in a compound. Because energy is proportional to frequency, and frequency determines color, you see only a few distinct colors in the spectrum of each element. This unique set of colors can then be used to determine which element produced the emission spectrum.

Light can also be produced by heating an object. You may have seen this before when a very hot object appears to glow. This method of emission produces a **continuous spectrum**, so you can see some amount of all frequencies within a particular range. An example of this is a hot stove coil that appears orange because of the particular set of frequencies it emits. The hot coil actually emits a whole range of wavelengths throughout the infrared and visible parts of the electromagnetic spectrum.

Safety Precautions

- Be very careful to not break the light bulb or spectral tubes to avoid broken glass.
- Do not plug in light sockets until the light bulbs are completely screwed in to avoid the risk of electric shock.
- Do not handle the spectrum tubes, power supply, or incandescent bulb with your bare hands after they have been used, since they will be extremely hot.

For this activity you will need the following:

- Incandescent light bulb
- Socket and plug for an incandescent bulb
- Spectrometer
- Spectrum tubes (neon, oxygen, nitrogen, helium)
- Spectrum tube power supply
- Thermal glove
- Colored pencil set

For this activity you will work *in pairs*.

Procedure

Step 1: First you are going to observe the emission spectrum for a spectral tube. Start with the tube of oxygen.

Step 2: Make sure that the spectrum tube power supply is *off* and insert the tube. Once the tube is inserted, turn the power supply *on*. The tube should begin to glow.

Step 3: Turn off the room lights so you only see light from the glowing tube.

Step 4: Look through the spectrometer toward the tube, as shown in Figure 12.2. Line the slit in the spectrometer up directly with the tube. Then look to the right of the slit inside the spectrometer. You should see a series of vertical lines of different colors at certain points along the wavelength scale that is at the bottom of your field of view.

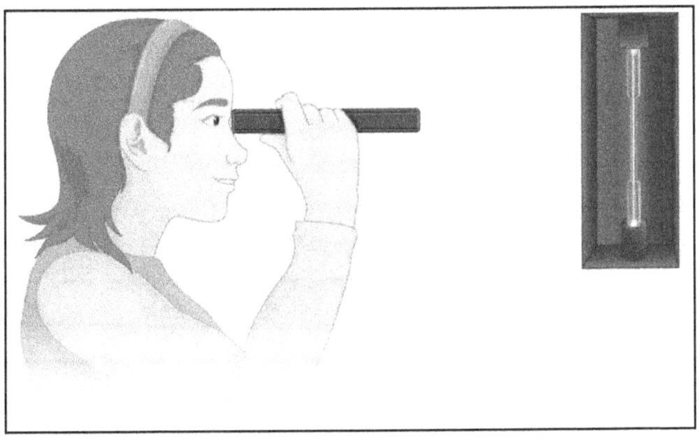

Figure 12.2: *When you look at the emission from a spectral tube through a spectrometer, it separates the light in the emission spectrum into its distinct frequencies.*

Step 5: Use your colored pencils to reproduce the spectrum you see through the spectrometer on Figure 12.3 in the *Data and Observations* section. Your emission spectra should match those shown in http://chemistry.bd.psu.edu/jircitano/periodic4.html.

Step 6: Your teacher has set up different stations where you can observe the spectra of several gases. Listen to your teacher carefully as he/she describes the gases and instructs how to move through the stations.

Step 7: Now, observe the spectrum for an incandescent bulb. Record the spectrum you see on the blank spectrum in Figure 12.3 in the *Data and Observations* section.

Data and Observations

Observed Emission Spectra

Oxygen Emission Spectrum

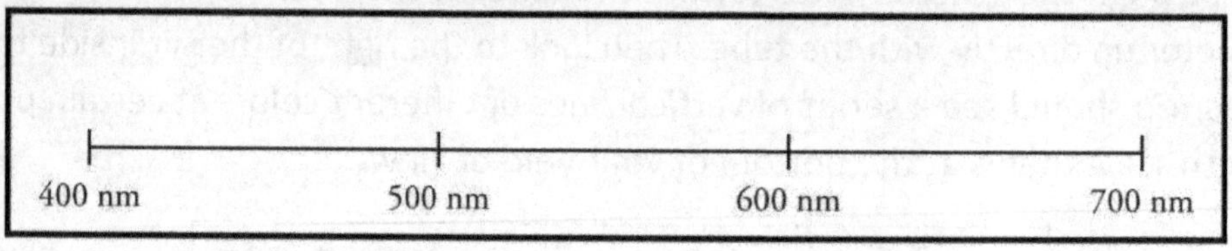

Nitrogen Emission Spectrum

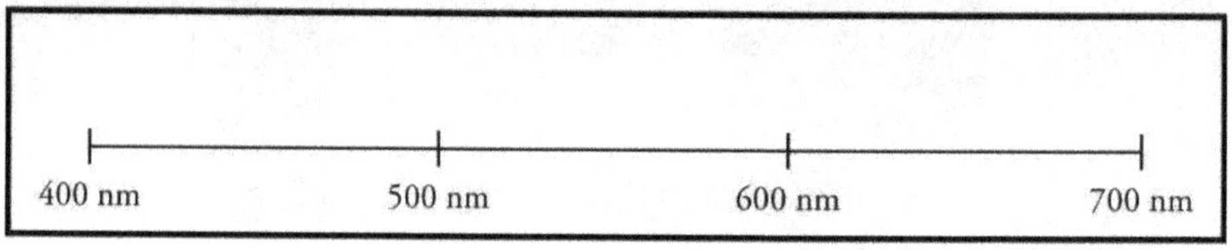

Helium Emission Spectrum

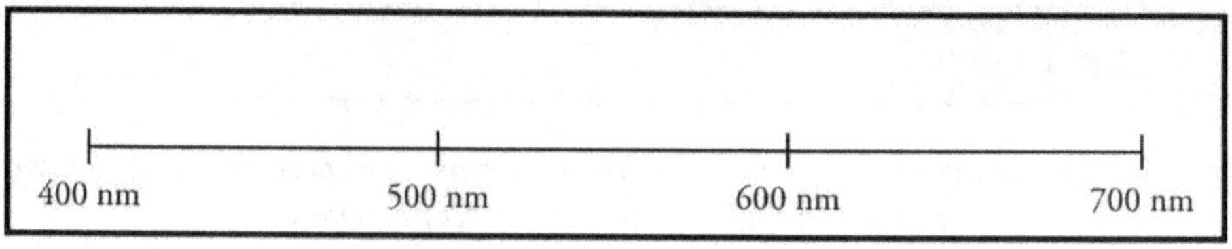

Neon Emission Spectrum

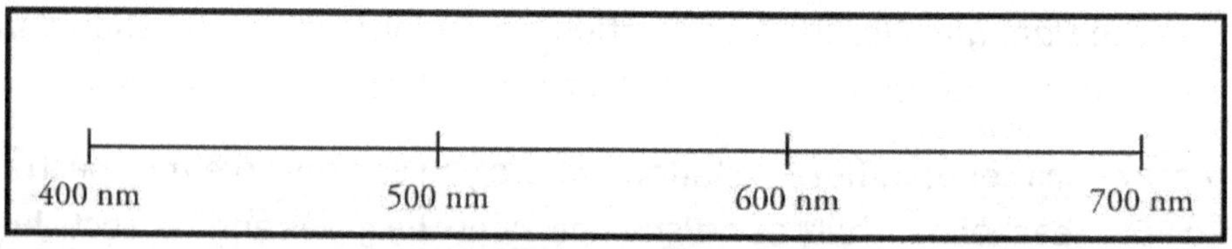

Incandescent Bulb Emission Spectrum

Figure 12.3: *The empty spectra should be filled in based on the spectra you observe using the spectral tubes or the incandescent bulb and spectrometer. Draw the lines spaced as you see them through the spectrometer, and use colored pencils to draw the lines with appropriate colors.*

Assessments

1. A spectrometer shows a bright line at a wavelength of 560 nm on an emission spectrum. TEKS (7)(C), TEKS (8)(B)

    a. What is the frequency of the light in the bright line?

    b. What is the energy of the light in the bright line?

    c. What is the energy difference between the levels through which an electron moved to, form this line?

2. While looking at an emission spectrum for a gas tube containing nitrogen, a student observes the emission lines expected for nitrogen, as well as some additional lines. What could explain the presence of these lines in the emission spectrum? TEKS (8)(B)

3. Every object emits electromagnetic radiation based on its temperature. The peak wavelength of this radiation is given by

$$\lambda = \frac{b}{T} \cdot$$

Where *b* (Wien's Constant) $= 2.9 \times 10^{-3}\ \mathrm{m \cdot K}$, and *T* is the temperature of the object in kelvins. TEKS (7)(C), TEKS (8)(B)

    a. If a person's body temperature is 310 K, what is the peak wavelength of the radiation that they emit?

    b. Based on Figure 12.01, is this radiation in the visible part of the electromagnetic spectrum? If not, in what part of the spectrum is it contained?

    c. Would you expect the spectrum emitted by the person to be discrete or continuous? Why?

Activity 2: Mixing Colors TEKS (7)(C)

When you think about mixing colors, you need to differentiate between mixing light or pigment. Since we see things by detecting light with our eyes, if we see no color, it looks dark. To see something, you need to add light. If you just add one color of light—for example, by turning on a light bulb of a particular color—you will see just that color. If you then add a second color, you will see the mixture of those two light colors. If you add all of the colors together, you see white light. This is why color mixing with light is called **additive color mixing**, because adding colors together determines what you see. The primary light colors are red, green, and blue. If you combine two of these colors, you get one of the secondary light colors—yellow, cyan, or magenta. If you combine all three primary colors, you get white light.

On the other hand, when you mix pigments, such as paint or ink in a printer, you start with white paper. The paper appears white because it reflects all colors of light, and all colors of light mixed together appear white. As you add pigment, you are causing certain colors to be absorbed by the pigment. The color that the pigment appears is the color that is reflected by the pigment. For example, red paint looks red because it reflects red light and absorbs other colors. If you mix all of the primary pigment colors together, it absorbs all colors of light and reflects none, causing it to appear black. This is why color mixing with pigment is called **subtractive color mixing**: you take colors away from white light, which has all colors, to determine what you see. The primary pigment colors are yellow, cyan, and magenta. If you combine two you get one of the secondary light colors—red, blue, or green. If you combine all three, you get black.

Safety Precautions

- Be very careful not to break the light bulb to avoid broken glass.
- Do not plug in light sockets until the light bulbs are completely screwed in to avoid the risk of electric shock.

For this activity you will need the following:

- Red, green, and blue light bulbs
- Meter stick
- Three light bulb sockets with plugs for incandescent bulbs

For this activity you will work *in pairs.*

Procedure

Step 1: Start by plugging in the red light bulb. Next, turn off the overhead lights in the room so you see only red light.
Step 2: Next, plug in the green light bulb. Record the color you see in the area where the red and green circles overlap in Figure 12.4 in the *Data and Observations* section.
Step 3: Use the meter stick to create shadows where the light from the two bulbs overlaps. You should observe that each primary color is visible in the shadow where the other color is blocked.
Step 4: Now unplug the green light bulb and plug in the blue light bulb. In Figure 12.2.mixing, record the color you see in the area where the red and blue circles overlap.
Step 5: Use the meter stick to create shadows where the light from the two bulbs overlaps. You should observe that each primary color is visible in the shadow where the other color is blocked.
Step 6: Now unplug the red light bulb and plug in the green light bulb. In Figure 12.4, record the color you see in the area where the green and blue circles overlap.
Step 7: Use the meter stick to create shadows where the light from the two bulbs overlaps. You should observe that each primary color is visible in the shadow where the other color is blocked.
Step 8: Plug in the red light bulb so you can observe the mixing of all three primary light colors. In Figure 12.4, record the color you see in the area where all three circles overlap.
Step 9: Use the meter stick to create shadows where the light from the three bulbs overlaps. You should observe that each secondary color is visible in the shadow where one of the three colors is blocked.

Data and Observations

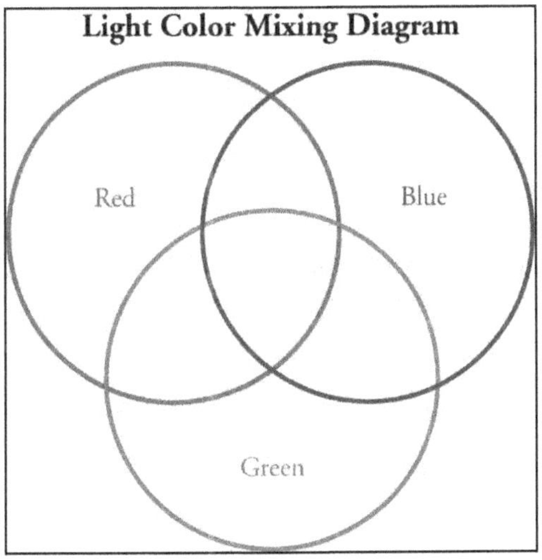

Figure 12.4*: Red, blue, and green are the primary colors of light. By mixing them in different configurations, you can produce secondary colors and white light. Fill in the blank parts of the diagram based on your observations to show what colors are produced by different configurations.*

Assessments

1. When you buy a color printer cartridge, the package often says that it is a CMY cartridge. What do the letters stand for, and why would a printer cartridge label carry these letters? TEKS (7)(C)
2. The parts of the visible spectrum that we call red, green, and blue are centered around the frequencies 650 nm, 510 nm, and 475 nm, respectively. Calculate the frequencies corresponding to these colors and show your work. TEKS (7)(C)
3. On the diagram below, complete a color wheel like the one in Activity 2 for pigment color mixing. The primary pigment colors should each go in one of the outer circles, the secondary colors should go in the overlap region of two circles, and in the center you should write the color you would get from mixing all three primary pigment colors. TEKS (7)(C)

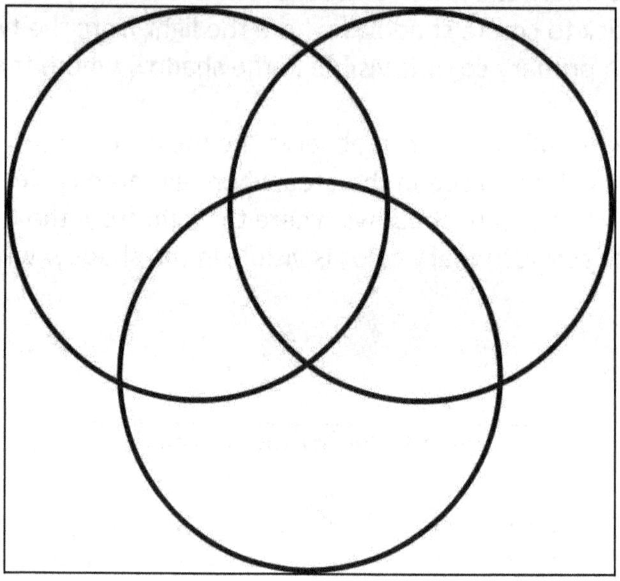

Figure 12.5: *Just as for light, there are three primary pigment colors and mixing them in different configurations yield secondary colors. Fill in the outer circles with the primary pigment colors and the overlap regions with the result of mixing those colors.*

Lab 13: Mirrors and Lenses

Modeling the interactions between mirrors, lenses, and light can assist in designing products, such as camera lenses. Mirrors and lenses are objects you interact with every day. Mirrors reflect light. Light strikes a mirror at the **angle of incidence** and bounces back at the **angle of reflection**. The **law of reflection** states that these two angles are equal. In contrast, light passes through a lens, but lenses often bend the path of the light. How the light behaves after traveling through the lens depends on the shape of the lens. A **convex lens** has a **focal point**, a spot where parallel rays of light passing through the lens will cross. Therefore, parallel rays converge at the focal point after passing through the lens. The distance to this point from the lens is known as the **focal length**. A **concave lens** causes light to diverge, meaning no ray of light passing through the lens will intersect another ray of light passing through the lens.

Ray diagrams are used to graphically describe the direction light travels before it strikes a medium (e.g., a mirror or lens) and where it goes afterward (e.g., reflection or refraction). In a ray diagram, arrowheads are used to denote the direction the light is traveling. Angles are measured relative to a line normal, or perpendicular, to the surface light strikes. These angles are used to refine the description of the path the light travels.

__In this lab you will learn:__

- how to investigate the behavior of light using ray diagrams;
- the relationship between a light's incident ray and its reflection;
- the relationship between lens type (convex or concave) and the behavior of refracted rays;
- how to determine the magnification of a lens.

Activity 1: Reflection with Plane Mirrors TEKS (7)(D)

When you see an object in a mirror, it is the result of light bouncing off the object and onto the mirror, which then reflects the light into your eye. Alternatively, the object may be a source of light itself, such as a light bulb. The ray of light hitting the mirror is referred to as the **ray of incidence.** The ray of light reflected back to you is referred to as the **ray of reflection**. If you trace the rays from the object to the mirror and back to your eye, it becomes more apparent why text in mirrors appears backward. In this activity, you will determine what path light takes as it travels to a mirror and is subsequently reflected.

Safety Precautions

- Mirrors are generally made of glass, so handle them carefully.
- Do not look directly into the laser; the light can damage your eye.
- Do not aim the laser at anyone's face.

For this activity you will need the following:

- Graphing paper (approximately four squares per inch)
- Colored pencils (four colors)
- Protractor
- Mirror
- Tape

For this activity you will work *in pairs*.

Procedure

Step 1: Turn a piece of graph paper lengthwise and tape the paper to the table.

Step 2: Place your mirror so that it is centered on the paper and perpendicular to it. Mark on your paper where the mirror is located.

Step 3: Several centimeters away from the mirror, draw a short arrow approximately three grid squares long, parallel to the mirror. (We use an arrow as the object viewed in the mirror so that the ends of the object can easily be identified.)

Step 4: Confirm you can see the arrow in the mirror.

Step 5: Set your laser so the beam passes over the arrowhead (an extreme, or end, point of your object), toward the mirror (Figure 13.1 ray diagram).

Step 6: The beam of light from the laser, passing over the arrowhead and striking the mirror, is the incident ray. Using a colored pencil, trace this ray on the paper from the arrowhead to the mirror.

Step 7: The beam of light passing away from the mirror is the reflected ray. Using a different colored pencil, trace the ray as it is reflected from the mirror.

Step 8: Repeat Steps 5 through 7 for the tail end of the arrow, using a different pair of colored pencils.

Step 9: Remove the mirror.

Step 10: On the incident rays, draw an arrowhead pointing toward and touching where the mirror was located. Label the rays.

Step 11: On the reflected rays, draw an arrowhead pointing away from the mirror. Label the rays.

Step 12: Continue the reflected rays through the mirror location with dashed lines and stop at the same distance away from the mirror as the arrow ends are located on the other side. This is where the image of the arrow would appear to be located in the mirror.

Step 13: Draw and label the normal line. This line is centered on and perpendicular to the mirror.

Step 14: Measure, record (in Table 13.1), and label the angles of incidence. These are the angles between the rays of incidence and the normal line.

Step 15: Measure, record (in Table 13.1), and label the angles of reflection. These are the angles between the rays of reflection and the normal line.

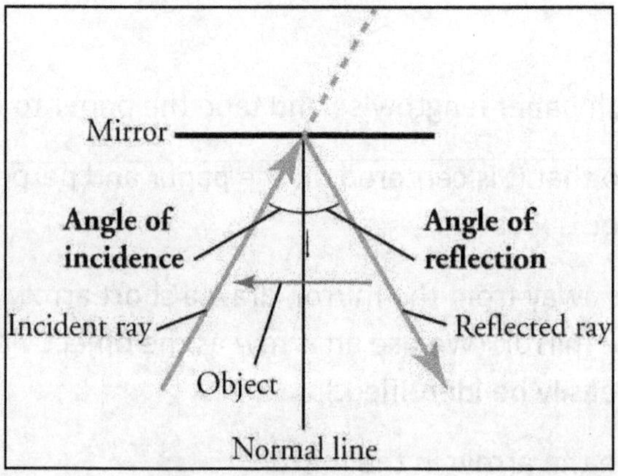

***Figure 13.1:** In this setup, the incident ray should pass over one end of the arrow. You should similarly draw the incident and reflected rays on your piece of paper, along with a normal line and measurements of the angles of incidence and reflection.*

Data and Observations

Table 13.1.Reflection from a Mirror

	Color of Incident Ray	Color of Reflected Ray	Angle of Incidence	Angle of Reflection	Observations Regarding Their Relationship	Observations Regarding Relationship to Normal Line
Head						
Tail						

Assessments

1. Draw and label the rays of incidence and rays of reflection, and the angles of incidence and angles of reflection, for the given scenario. TEKS (7)(D)

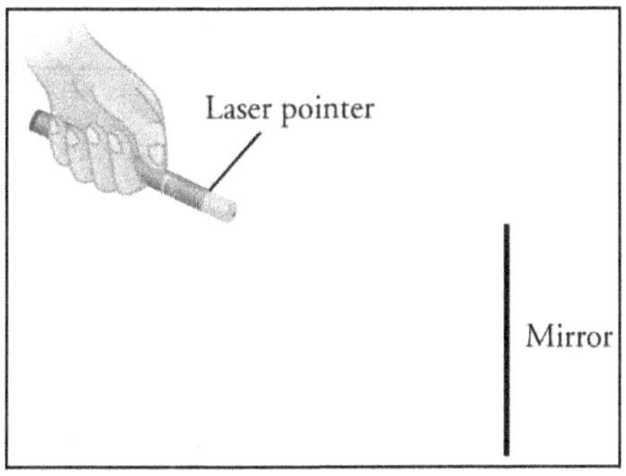

2. Use a ray diagram to determine whether Juan can see Melissa in the mirror, based on the diagram below. TEKS (7)(D)

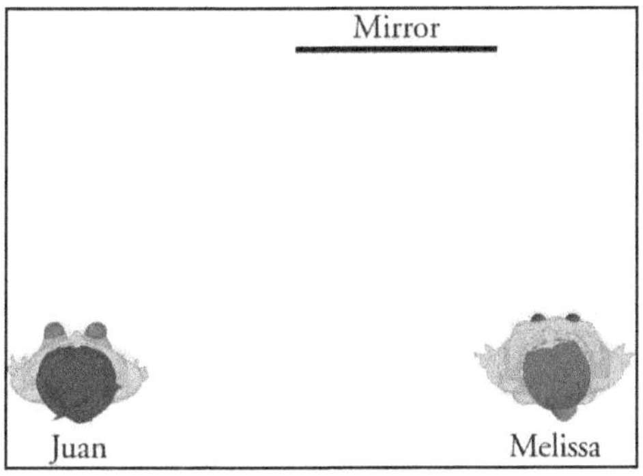

3. When driving, it is important that you are able to see everything around you when determining if it is safe to change lanes. Using the concepts covered in this activity, explain how a blind spot can occur on the driver's side if the mirrors are not angled correctly. TEKS (7)(D)

Activity 2: Refraction with Lenses TEKS (7)(D)

The ability of concave and convex lenses to refract light is the driving principle behind corrective lenses, such as those in eyeglasses. Ideally, the lens of the human eye focuses light on the retina, located at the back of the eye. The lens of a farsighted individual has a focal point that goes behind the retina, which requires a convex lens to bring the image forward. In contrast, a nearsighted person has a focal point ahead of the retina, which requires a concave lens to set the image back on the retina. In this activity, you will explore the effect lens shape has on the focal length.

Safety Precautions

- Lenses are generally made of glass, so handle them carefully.
- Do not look directly into the laser; the light can damage your eye.
- Do not aim the laser at anyone's face.

For this activity you will need the following:

- Colored pencils (four colors)
- Graphing paper (approximately four squares per inch)
- Laser
- Convex lens
- Concave lens
- Tape
- Ruler

For this activity you will work *in pairs*.

Procedure

Step 1: Tape together two sheets of graph paper along the short side of each piece. Gently tape the corners of the paper to the table. *Note*: You will need to be able to use the back of the pages as well in this activity.

Step 2: Draw a set of axes with an origin at the center of the taped pieces of paper.

Step 3: Place a convex lens so that it is centered on the origin and parallel to the short sides of the paper.

Step 4: Place the laser on the axis parallel to the long sides of the paper so that the beam travels down the axis through the lens. Adjust the position of your lens so that the beam travels straight through the lens and along the line on the other side. Trace the outline of the lens when you have it positioned correctly on the paper.

Step 5: Move the laser about 20 cm (approximately 8 in.) from the lens along the axis, then move it up so that the beam still passes through the lens. The light will pass through the axis on the other side of the lens. Trace the path of the laser with a colored pencil from the device, through the lens, and beyond where it passes through the axis.

Step 6: Move the laser the same distance below the line, still 20 cm (approximately 8 in.) from the lens. Trace the path of the laser with a different colored pencil from the device, through the lens, and beyond where it passes through the axis.

Step 7: The two lines that you drew should meet on the axis, though they might be slightly off due to defects in the lens. Mark the point where they meet (or the midpoint of the two places the lines cross the axis) and label it the focal point (Figure 13.6). Measure the distance from the origin (on which the lens is centered) to the focal point. This is the focal distance, which you will need later in this activity. Record the focal distance.

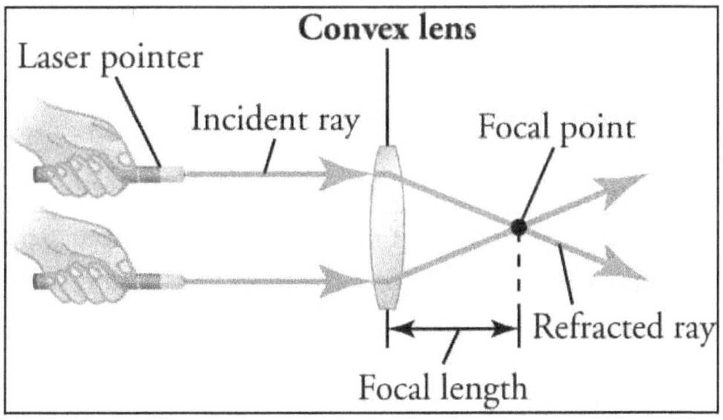

Figure 13.6: *The result of shining two parallel beams of light on alternate ends of a convex lens. The two beams cross on the far side of the lens, which is the focal point of the lens. The distance from the center of the lens to the focal point is the focal length.*

Step 8: Place a concave lens so it is centered on the origin and parallel to the short sides of the paper.

Step 9: Place the laser on the axis parallel to the long sides of the paper so that the beam travels down the axis through the lens. The beam should travel straight through the lens and along the line on the other side. Adjust the position of your lens so that this happens. Trace the outline of the lens when you have it positioned correctly.

Step 10: Move the laser about 20 cm (approximately 8 in.) from the lens along the axis, then move it up so that the beam still passes through the lens. The light will lead away from the axis after it passes through the lens (Figure 13.7). Trace the path of the laser with a different colored pencil from the device through the lens, and beyond to the edge of the paper.

Step 11: Move the laser the same distance below the line, still 20 cm (approximately 8 in.) from the lens. Trace the path of the laser with a different colored pencil from the device through the lens, and beyond to the edge of the paper.

Step 12: Observe and record the differences in refracted rays between the convex and concave lenses.

Step 13: On the other side of your paper, repeat Steps 1 through 4.

Step 14: About 15 cm (approximately 6 in.) from the lens, draw an arrow that starts on the axis parallel to the long side of the page, running perpendicular to that axis, and stops three grid squares above the axis, with the head of the arrow drawn at the end farther from the axis. Record (in Table.13.2) the length of your arrow.

Step 15: Place the laser so that the beam originates at the arrowhead and is parallel to the horizontal axis. Trace the beam from the laser, through the end of the arrow, to the lens (the incident ray), and along its path on the other side of the lens (the refracted ray), until you reach the end of the paper.

Step 16: Place a dot on the horizontal axis between the laser and the lens at the focal length from the lens, based on the distance you found in Step 7.

Step 17: Keeping the laser so that the beam originates at the arrowhead, rotate it so the beam also passes through the focal point. Trace the beam from the laser through the end of the arrow, to the lens (the incident ray), and along its path on the other side of the lens (the refracted ray), until you cross the first incident ray.

Step 18: At the point where the rays intersect, draw an upside-down arrowhead.

Step 19: Draw a line perpendicular to the horizontal axis from the arrowhead you drew in Step 18 to the axis. This arrow is the image of your object. Measure and record (in Table.13.2) the length of the arrow.

Step 20: Calculate the magnification *M* of the image. The magnification of an object can be found in two ways. The first is to divide the length of the image by the length of the object. The second is by using the thin lens equation, which is

$$M = \frac{\text{focal length}}{\text{focal length} - \text{distance of the object from the lens}}.$$

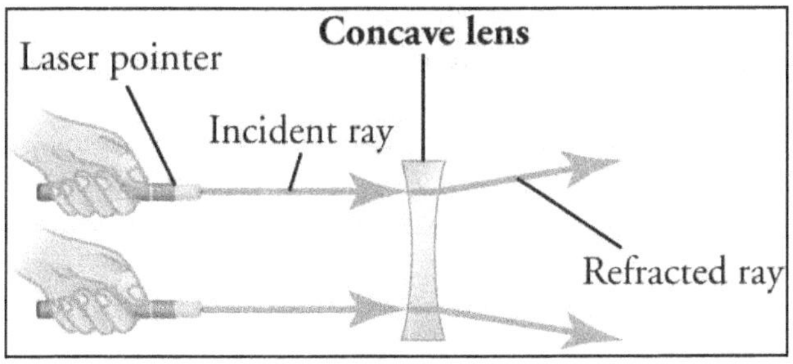

Figure 13.7: *The result of shining two parallel beams of light on alternate ends of a concave lens. The two beams diverge on the far side of the lens.*

Data and Observations

Focal length:

Observations regarding the incident and refracted rays of the convex lens:

Observations regarding the incident and refracted rays of the concave lens:

Table 13.2.Magnification

Length of Object	Length of the Image	Magnification

Assessments

1. What type of lens might you use to maximize the light emitted by a light bulb? Explain your answer, using the properties of the lens type you selected. TEKS (7)(D)
2. You are given the task of finding a lens to magnify the image of an object to twice the object's size. What type of lens would you use and what would the focal length of the lens need to be? What other considerations might you need to make? TEKS (7)(D)
3. The eye is capable of changing its focal length relative to the object focused on, a process referred to as accommodation. To look at an object closer to you, should your lens become more or less convex? Justify your answer. TEKS (7)(D)

Lab 14: Refraction through Different Media

Have you ever tried to grab an object from the bottom of a swimming pool? Before you put your head under water, the object appears to be in one place. After your eyes go below the surface, the object appears to be in another place. The location of the object appears to change because the light entering your eyes **refracts**, or changes direction, as it travels between air and water. Refraction occurs because light travels more slowly in water than in air.

Light travels fastest in a vacuum, at a speed of 3.00×10^8 m/s. Light travels more slowly through matter because it interacts with the atoms that make up the matter. The amount of interaction that occurs, and thus the speed at which light travels, depends on the atomic makeup of the matter, or medium.

Light interacts more with liquid water than with air, and therefore travels more slowly through liquid water than air. When light travels across an air-water boundary, its speed changes, which causes a change in its direction of travel.

To understand why a change in speed causes light to change direction, or bend, let us consider an analogous situation in which a lawnmower slows as it moves from a sidewalk to grass. If the lawnmower enters the grass at an incident angle of θ_1, the right front wheel slows before the left front wheel, causing the lawnmower to turn toward the *normal*, a line perpendicular to the boundary between the grass and sidewalk (Figure 14.1 (a)). Because the lawnmower turns, angle θ_2 is less than angle θ_1.

If the lawnmower leaves the grass at angle θ_1, the right front wheel speeds up before the left front wheel, causing the lawnmower to turn away from the normal (Figure 14.1 (b)). As a result, angle θ_2 is greater than angle θ_1.

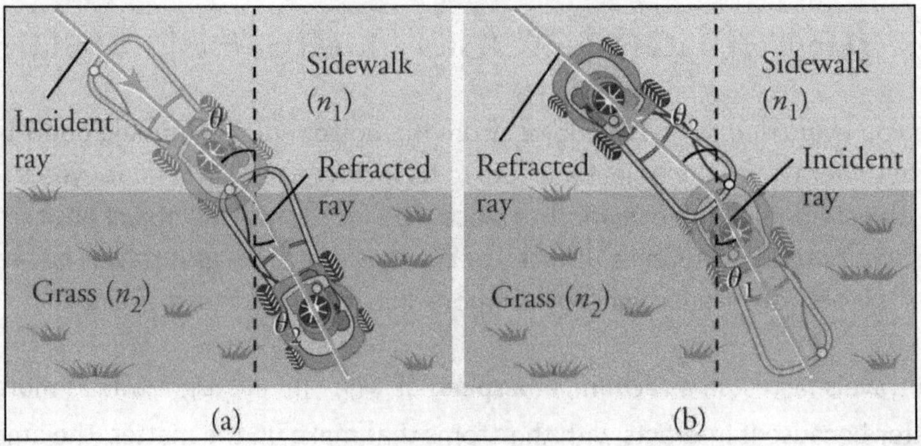

Figure 14.1: *(a) A lawnmower slows as it moves from the sidewalk* (n_1) *to grass* (n_2) *at an angle of* θ_1*. The right wheel slows before the left wheel, causing the mower to turn so that* θ_2 *is less than* θ_1*. (b) The mower speeds up as it moves from the grass to the sidewalk, so* θ_2 *is greater than* θ_1*.*

The index of refraction (*n*) is the ratio of the speed of light in a vacuum (*c*) to its speed in a given type of matter (*v*),

$$n = \frac{c}{v}.$$

Since the index of refraction is a ratio, it has no units. If the index of refraction of one substance is known, the index of refraction of another can be calculated using Snell's law. Snell's law can also be used to calculate the speed of light in a particular medium. We write it as

$$\frac{n_1}{n_2} = \frac{\sin\theta_2}{\sin\theta_1} = \frac{v_2}{v_1}.$$

In the lawnmower analogy above, a simple linear boundary exists between the sidewalk and grass. In real life, boundaries are often less clear and/or much more complex. Refraction across complex boundaries can produce some interesting optical effects. For example, multiple images may be produced if an object is viewed through more than one medium that refracts light (Figure 14.2 (a)).

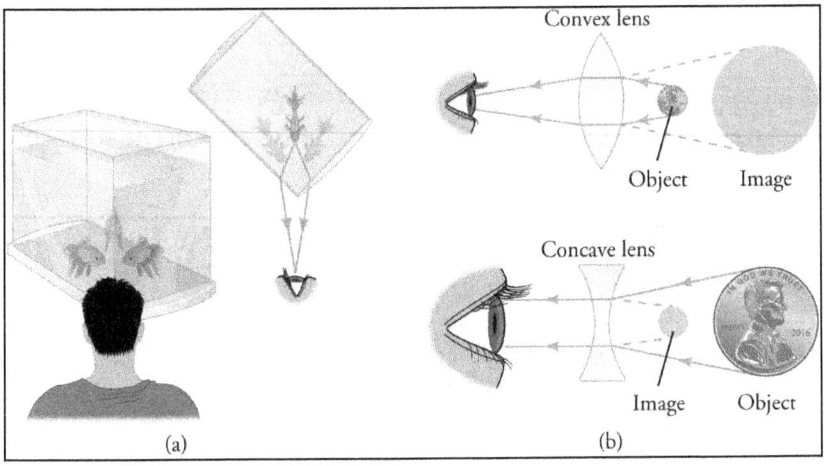

Figure 14.2: *(a) Multiple images may be produced if an object is viewed through more than one refractive surface. (b) Curved surfaces can change the apparent size and location of an image.*

If refraction occurs through a **convex lens**—one that bulges in the middle—the light waves diverge before reaching the boundary, causing the object to appear larger and farther away than it really is (Figure 14.2 (b)). If refraction occurs through a **concave lens**—one that is thinner in the middle and thicker on the edges—the light waves converge before reaching the boundary, causing the object to appear smaller and closer than it really is.

In this lab you will observe refraction through different materials and through boundaries with different shapes. In Activity 1, you will observe optical effects produced by refraction through water and determine the cause of these effects. In Activity 2, you will compare the index of refraction of water to that of vegetable oil and corn syrup. In Activity 3, you will determine the index of refraction of acrylic.

<u>In this lab you will learn:</u>

- how refraction can shrink, magnify, or distort an image;
- how refraction can cause multiple images to appear, or cause an object to disappear;
- how to assess the relative index of refraction of two different materials;
- how to calculate the index of refraction of a material from the angle of incidence and the angle of refraction.

Activity 1: Refraction Through Water TEKS (7)(D)

A drinking glass is an everyday object that you may take for granted. But this simple object varies in thickness and shape. Refraction through these different surfaces can distort an image in interesting ways (Figure 14.3).

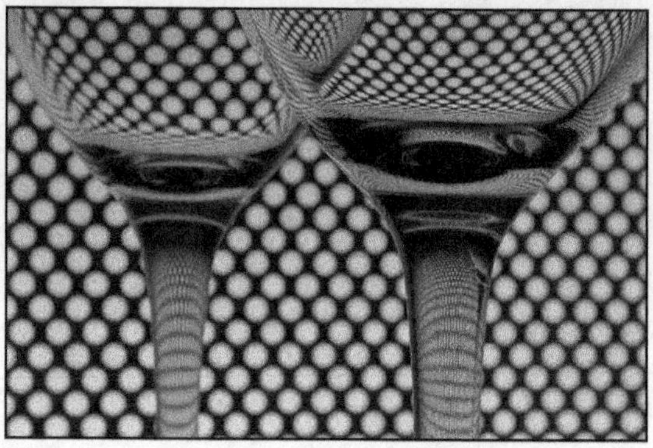

Figure 14.3: *These drinking glasses can produce complex refraction patterns.*

Looking into a glass filled with water, the light must not only pass through the water, it must also pass through the glass. Glass and water have different indexes of refraction, so the light bends twice before reaching your eye: Once while moving from water to glass, and again while moving from glass to air. In this activity, you will explore some optical illusions that occur when light passes through water and glass and determine the cause of the illusion.

Safety Precautions

- Use caution with glass objects, and alert your teacher if anything breaks.

For this activity you will need the following:

- Clear glass cup
- Clear rectangular water container
- Penny

For this activity you will work *in pairs*.

Procedure

Step 1: Make money grow and shrink.

- Fill a glass with water. Hold a penny so that it is half-submerged, about two-thirds of the way back in the glass. Look at the penny's flat side so you can see both the part above and below the water, as shown in Figure 14.4 (a)). It may be easier to see the penny if you set the glass on white paper. What do you see? Does the part of the penny below water look larger or smaller than the part above? Does it look closer or farther away? Is it inverted? Record your results in Table 14.1.

- Partly fill a clear rectangular container with water. Place an empty cup in the container; hold it down if it floats. Place a penny behind the empty cup so that it is half-submerged. Look down at the penny, so you can see it through the backside of the glass. See Figure 14.4 (b)). Make sure you can see both the part above and below the water. What do you see? Record your results in Table 14.1.

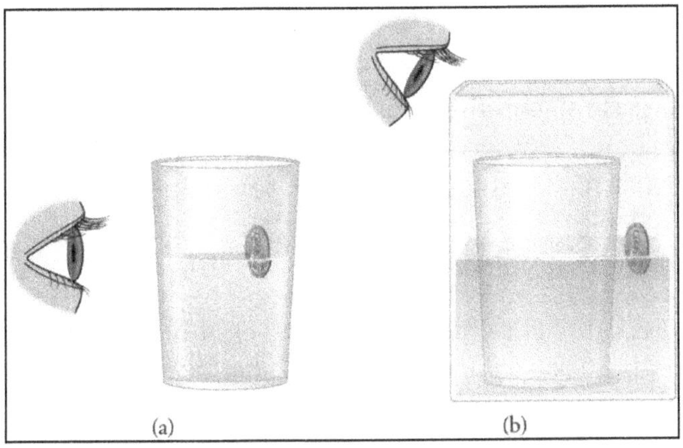

Figure 14.4: *(a) Partly submerge a penny in a glass of water and view it from the side. (b) Partly submerge a penny in a rectangular container and view it from behind the backside of an empty cup.*

- Why does the penny grow and shrink? When the partly-submerged penny is inside the glass, you view it through a boundary that curves *outward*. The outward curvature causes rays of light to diverge as they travel from the penny to your eye. Thus, the glass acts as a **convex** lens that magnifies the image of the penny and makes it appear farther away.

- When the partly-submerged penny is behind the glass, you view it through a boundary that curves *inward*. The inward curvature causes rays of light to converge as they travel from the penny to your eye. Thus, the glass acts as a **concave** lens that shrinks the image of the penny and makes it appear closer.

Step 2: Make money multiply.

- Partly fill a rectangular container with water. Look at the penny through a corner of the container, so you can see the top surface and both sides. How many pennies do you see? Move the penny around. How much money can you make?

Step 3: Make money disappear.

- Set an empty glass on top of a penny. Make sure both the penny and the bottom of the glass are dry. Look through the side of the glass. Can you see the penny?
- Fill the glass with water, looking at the penny through the side as you fill the glass. Does the penny disappear?
- Now, wet the penny and put it back under the glass. Can you see it now?
- Why is the penny sometimes visible and sometimes not?
- Glass has an index of refraction of about 1.5 and air has an index of refraction of about 1.0. When the glass is empty, light bends upon entering the glass bottom, but bends back again upon entering the air in the glass. Its path to your eye is nearly straight, so you can see the penny (Figure 14.5 (a)).
- When the glass is full and the penny is dry, light bends upon entering the glass bottom, but it doesn't bend back as much upon entering the water because the refractive index of water (1.33) is higher than the index of refraction of air. Since the light doesn't bend back as much, light travels upward toward the top of the glass. If you look straight down, you will be able to see the penny (Figure 14.5 (b)).
- When the glass is full and the penny is wet, light bends upon traveling from water to glass, but bends back upon traveling from glass to water. Its path to your eye is once again nearly straight so you can see the penny (Figure 14.5 (c)).

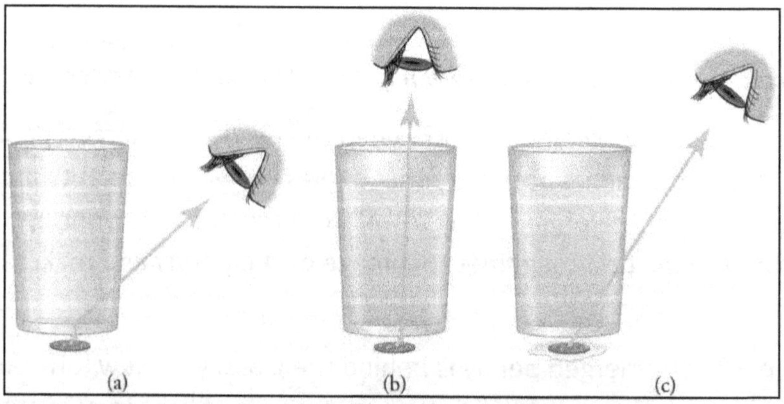

***Figure 14.5:** (a) Light bends upon entering the glass bottom, but bends back again upon entering the air in the glass. (b) Light bends upon entering the glass bottom, but it doesn't bend back as much upon entering the water because the refractive index of water is higher than that of air. (c) Light bends upon traveling from water to glass, but bends back upon traveling from glass to water.*

Data and Observations

Table.14.1: Money that Grows and Shrinks

Experiment	Boundary Shape (Convex or Concave)	Is the Image Under Water Inverted?	Does the Image Under Water Appear Magnified or Shrunken?	Does the Image Under Water Appear Closer or Farther Away?
Penny in the cup				
Penny behind the cup				

Assessments

1. A butterfly viewed through glass appears larger and farther away. Is the surface of the glass convex, concave, or flat? Explain. TEKS (7)(D)
2. When light crosses a boundary into a medium with a higher index of refraction, its speed .... (increases/decreases/remains the same). TEKS (7)(D)
3. Describe a situation in which refraction can cause multiple images to appear. TEKS (7)(D)

Activity 2: Refraction Through Different Liquids TEKS (7)(D)

Light not only travels through water, it also travels through other clear liquids, such as oil. In this activity you will compare the relative index of refraction of three liquids: oil, water, and corn syrup.

Safety Precautions

- Use caution with glass objects, and alert your teacher if anything breaks.

For this activity you will need the following:

- Clear plastic or glass cup
- Straw
- Vegetable oil
- Corn syrup

For this activity you will work *in pairs*.

Procedure

Step 1: Observe a straw in a cup.

- Fill a transparent cup with water. Place the straw in the cup so that it is vertical and move the straw in a slow circle. Look at the interface between air and water. What do you see? As you move the straw, does it appear broken, enlarged, shifted to the right, or shifted to the left?

Step 2: Compare the relative refraction of water, oil, and corn syrup.

- Empty the cup. Pour in corn syrup to about one-third full. Carefully pour a water layer over the corn syrup layer, to about two-thirds full. Carefully pour vegetable oil over the water. Leave a bit of room at the top, so the liquid doesn't spill over.

Step 3: Determine the relative index of refraction for each boundary.

- Place a straw vertically in the cup toward the left through all three layers. For each layer, note whether the straw appears shifted to the left or to the right, relative to the layer above. Record your results in Table.14.2.

Data and Observations

What do you see when you move the straw around the cup? Record your results below.

Table 14.2 Relative Index of Refraction

Layer	Shift Relative to the Layer Above (left or right)	Relative Index of Refraction*
Vegetable oil		$n_{\text{vegetable oil}}$ n_{air}
Water		N_{water} $n_{\text{vegetable oil}}$
Corn syrup		$n_{\text{corn syrup}}$ n_{water}

* When the straw is on the left side of the cup, a shift to the left indicates a higher value in the index of refraction, and a shift to the right indicates a lower index of refraction. Insert a greater than (>), less than (<), or equal to (=) sign to indicate what shift occurs.

Assessments

1. A student observes that the image of a straw shifts more at an ethanol-air boundary than a water-air boundary. Which has a higher index of refraction, ethanol or water? Explain. TEKS (7)(D)
2. A student notices that a sugar water solution magnifies a penny more than pure water. Which liquid has a higher index of refraction? Explain. TEKS (7)(D)
3. When light crosses a boundary from acrylic to water, the angle of refraction is greater than the angle of incidence. Which has a higher index of refraction, acrylic or water? Explain. TEKS (7)(D)

Activity 3: Refraction Through Clear Solids TEKS (7)(D)

In Activity 2, you compared the relative index of refraction of different liquids. In this experiment you will use a laser to determine the index of refraction of a clear solid.

Safety Precautions

- Do not point the laser into anyone's eye.

For this activity you will need the following:

- Block of acrylic or other clear substance
- Protractor
- Paper
- Pencil
- Laser

For this activity you will work *in pairs.*

Procedure

Step 1: Set up the experiment as shown in Figure 14.6.

- Draw two perpendicular lines on a sheet of paper. Label one line *normal* and the other line *boundary*.
- Using a protractor, draw lines at 10°, 20°, 30°, and 40° to the left of the normal.
- Place an acrylic block so that the upper edge is flush with the boundary line. The center of the block should rest about where the boundary line intersects the normal line. Draw a line around the block.

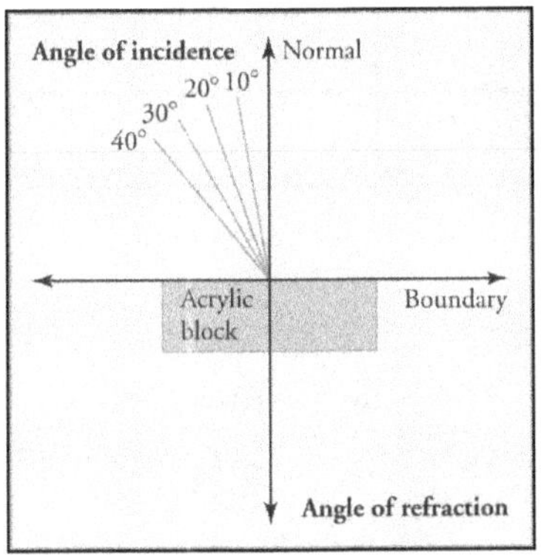

***Figure 14.6**: Setup for Activity 3, refraction through a solid.*

Step 2: Make the measurements.

- Set the laser on the paper and adjust its position so that the beam travels along the 10° line toward the block. You should be able to see the laser through the acrylic and after it exits the block. If the block gets shifted during the measurement, move it back.
- Draw a line indicating the direction the beam travels after it exits the block.
- Repeat this measurement for the 20°, 30°, and 40° lines.
- Turn off the laser and remove the block from the paper.

Step 3: Determine the angle of refraction.

- Continue the line for each measurement backward, to the point where the beam exited the block. Draw a line from this line to the point of intersection between the normal and boundary lines.
- Use a protractor to determine the angle of refraction for each measurement. Record your results in Table 14.3.
- Calculate the index of refraction of acrylic (n_2) and record your results in the last column of Table 14.3.
- Calculate the average index of refraction of acrylic (n_2) and the speed of light in acrylic and record your results in Table 14.4.

Data and Observations

Table 14.3: Index of Refraction of a Solid

Angle of Incidence (θ_1)	Angle of Refraction (θ_2)	$\sin\theta_1$	$\sin\theta_2$	$n_2 = \frac{\sin\theta_1}{\sin\theta_2}$
10°				
20°				
30°				
40°				

Note—According to Snell's law,

$$\frac{n_1}{n_2} = \frac{\sin\theta_2}{\sin\theta_1}\ .$$

The index of refraction of air is one $(n_1 = 1)$. Thus,

$$\frac{1}{n_2} = \frac{\sin\theta_2}{\sin\theta_1}\ .$$

Rearranging,

$$n_2 = \frac{\sin\theta_1}{\sin\theta_2}\ .$$

Table 14.4.Speed of Light in Acrylic

Average index of refraction in acrylic (n_{2ave})	
Speed of light in acrylic (v)*	

Note: $n_{2ave} = \frac{c}{v}$, where $c = \text{speed of light in air} = 3\times10^8$ m/s, and v = speed of light in acrylic.

Rearranging,

$$v = \frac{3\times10^8}{n_{2ave}}\ \text{m/s}\ .$$

Assessments

1. TEKS (7)(D)

   Light travels from Medium *A* to Medium *B* as shown in the image below. Which medium has the higher index of refraction? Explain.

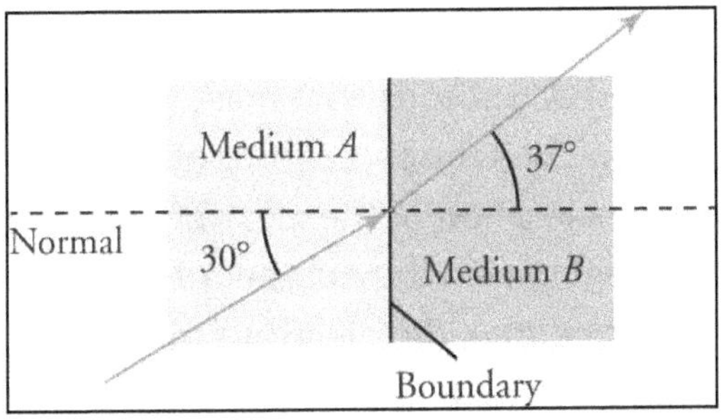

2. Cubic zirconia has an index of refraction of 2.15 and air has an index of refraction of 1.00. If light enters cubic zirconia from air at an angle of 30°, what will the angle of refraction be? TEKS (7)(D)

3. Light enters diamond from air at an incident angle of 45°. The angle of refraction is 17°. What is the index of refraction of diamond? TEKS (7)(D)

Lab 15: Electric Charge

Matter usually contains equal numbers of protons (positively-charged particles) and electrons (negatively-charged particles). So, when an object is positively charged, it has lost electrons. If it is negatively charged, it has gained extra electrons.

Electrons are more mobile than protons. The mass of an electron is over a thousand times smaller than that of a proton. Thus the same force applied to a free electron and a free proton results in a much higher acceleration of the electron. Electrons move when they are not balanced across two locations. Imagine that the number of electrons is building up in one object. If another object is brought nearby, the electrons may *jump* between the objects. The electrons will be moving until the electrostatic potential is equalized.

A net charge, or imbalance of charge, is called an **electrostatic charge** or **static charge**. In this lab, you will build a simple device to detect charges on objects, and investigate how objects can become charged. You will also examine how charge can be stored.

In this lab you will learn:

- how to use the idea that similar charges repel and opposite charges attract to build a device for detecting charge;
- how to charge objects using friction, conduction, and induction;
- how to store charge using a capacitor.

Activity 1: Charge TEKS (5)(C)

"Opposites attract" is a true statement when dealing with electricity. If two objects have opposite charges, they are attracted to one another; and if they have the same charge, they move apart. An **electroscope** is a device used to detect static charges. It consists of two small and very light pieces of metal, called leaves, suspended from a wire. The top of the wire acts as a charge collector and the charge moves down the wire to both metal leaves. Any excess charge will spread evenly across the surface of the leaves. This will cause the leaves to repel each other because both leaves carry charge of the same sign. We can use this property to detect the presence of a charge imbalance.

Note—In this experiment, you will build an electroscope. This experiment works better in dry weather, and is more difficult to perform when it's humid. Placing a ball of foil on top of the loop can sometimes increase the sensitivity of the electroscope. Also, the oils on your skin can prevent the electroscope from working, so handle the foil as little as possible. If you need to clean the foil, use a cotton swab and rubbing alcohol to clean the surface.

Safety Precautions

- Use caution when working with sharp objects, including thumbtacks and scissors.

For this activity you will need the following:

- Clean, empty 16-ounce plastic bottle
- Metal paper clip
- Thumbtack
- Piece of copper or silver foil
- Scissors
- Ruler
- Balloon
- Wool

For this activity you will work *in pairs*.

Procedure

Step 1: Cut a strip of foil that is about 0.25 inches (or, about 0.6 cm) wide and 4.5 inches (or, about 10 cm) long. On each end, fold over the last 0.25 inches (or, about 0.6 cm) of foil and crease. Fold the piece of foil in half so that it is 2 inches (or, about 5 cm) long on each side. See Figure 15.1.

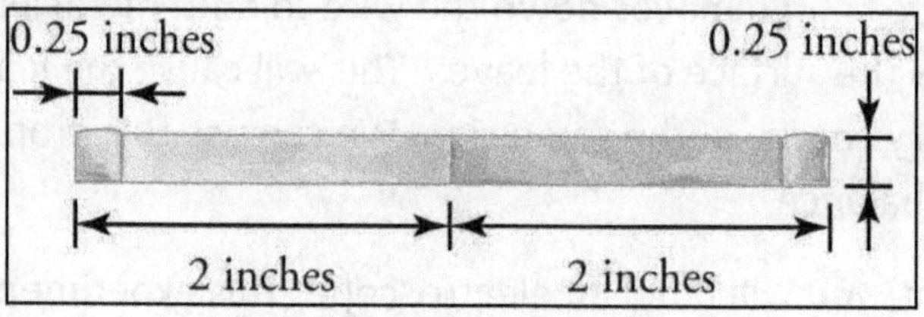

***Figure 15.1**: The piece of foil that will become the electroscope leaves should be approximately 4 inches (or, 10 cm) long.*

Step 2: Push a hole through the center of the bottle cap with the thumbtack.

Step 3: Straighten the paper clip and bend the last half inch into a *U* shape.

Step 4: Push the straight portion of the paper clip through the hole in the bottle cap so that the *U* end is below the cap. Place the foil strip onto the paper clip so the center fold rests on the *U* bend.

Step 5: Screw the cap onto the bottle. Slide the paper clip to adjust the height so the foil hangs approximately 1 inch into the bottle but does not touch the bottle. Bend the top of the paper clip into a loop so that it is fixed in place (Figure 15.2).

Step 6: Lightly rub an inflated balloon with a piece of wool for 30 seconds. Bring the balloon close to the top of the electroscope and observe the leaves.

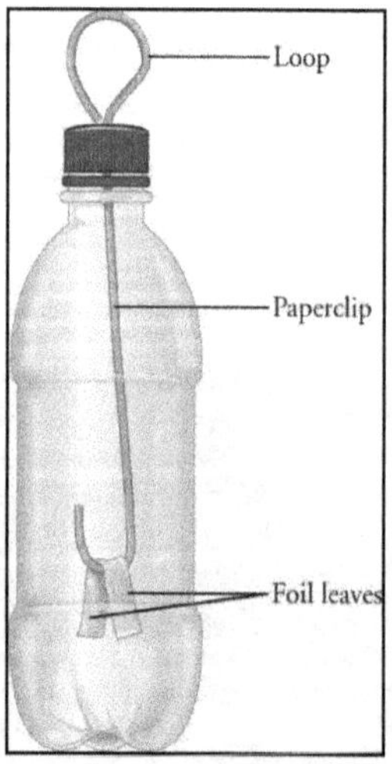

***Figure 15.2:** The foil leaves of the electroscope hang down straight due to gravity, but push outward in the presence of a charged object.*

Data and Observations

Table.15.1 Electroscope Observations

What observations did you make while constructing the electroscope? Did you encounter any challenges?	
How far apart are the leaves (not separated, a little separated, or a lot separated) when the balloon is:	
1 foot away from the foil leaves?	
2 inches away from the foil leaves?	
1 inch away from the foil leaves?	

Assessments

1. What is likely happening to the electrons when a balloon is rubbed against wool? TEKS (5)(C)
2. Does the distance between the electroscope and the charged object influence the force pushing the leaves apart? Explain. TEKS (5)(C)

Activity 2: Friction, Conduction, and Induction TEKS (5)(E)

Many people are familiar with static charge because it has the annoying property of generating stinging shocks. If you walk across a carpet on a cold, dry day and then touch a metal doorknob, you might receive a small jolt. This experience demonstrates two ways that charge can be transferred. When two different materials rub across one another, electrons are often transferred. Charge can be transferred by friction, which happens when you run a plastic comb through your hair, rub a balloon with wool, or walk across a carpet. One material has a greater affinity for electrons and accepts them from the other material. The total charge is conserved, but it is now distributed unevenly between two objects.

Charge can move through many objects. However, **conductors** are materials through which electrons can easily move. When a charged object touches a conductor, electrons rapidly flow from one object to another to reduce the imbalance. We experience this when we receive electric shocks from metal objects after walking across a rug or carpet. When you build up a charge through friction and touch a piece of metal, you are a charged object touching a conductor, and you experience the rapid flow of electrons as a shock.

Transferring charge through friction, as well as through conduction, usually requires physical contact, as this contact allows the charges to move. Another method of transferring charge, called **induction**, requires objects to be close to each other but not touching. Opposite charges attract and like charges repel even through empty space. Therefore, one charged object can influence the charge on another object even when no electrons are exchanged between them.

In this activity, you will use an electroscope to investigate various objects that are charged by friction. Bringing charged objects near the electroscope uses induction to concentrate a charge on the foil leaves. Touching the electroscope while the leaves are charged provides an opportunity for the charge to leave the leaves through conduction, but only if the material is a conductor or a previously charged insulator. In this activity, you will investigate transferring charge through friction, induction, and conduction, and determine if materials are conductors or **insulators**, that is, materials that do not allow electrons to easily flow.

For this activity you will need the following:

- Electroscope
- Two balloons
- Piece of wool
- Hard plastic comb or rod
- Glass rod
- Wooden rod
- Small cm ruler

For this activity you will work *in pairs*.

Procedure

Step 1: Rub a balloon with wool for 30 seconds. Bring the balloon close enough to an electroscope to cause the foil leaves to move apart. Keep the balloon in this position while your partner measures the distance between the very tips of the leaves, in centimeters. Also make a mental note of the position of this balloon.

Step 2: Rub a second balloon with wool for 30 seconds. Bring the second balloon near the electroscope and record any changes you observe. Determine if the relative position of the two balloons influences the electroscope by moving the second balloon. Keep the distance between the balloon and the electroscope constant, but move in an arc so the distance between the balloons changes (Figure 15.3). Record your observations in Table 15.2. Do the leaves move farther apart, closer together, or remain unchanged?

Step 3: Recharge one balloon by rubbing it again with wool. Place the balloon in the same position it was in during Step 1. Bring the wool close to the electroscope and record your observations in Table 15.2.

Step 4: Rub a plastic comb or plastic rod with the piece of wool for 30 seconds. With the balloon still positioned near the electroscope, bring the plastic object near the electroscope and record any observations in Table 15.2.

Step 5: Repeat the procedure for the previous step using the glass rod rather than the plastic object. Again, place the charged balloon near the electroscope, and observe the movements of the foil leaves as the glass rod is brought close to the electroscope.

Step 6: Bring any charged object close enough to the electroscope to cause the foil leaves to move. Touch the metal portion of the electroscope with a wooden rod, and record any observations in Table 15.2.

Step 7: Again, bring any charged object close enough to the electroscope to cause the foil leaves to deflect. Touch the metal portion of the electroscope with your bare hand and record any observations in Table 15.2.

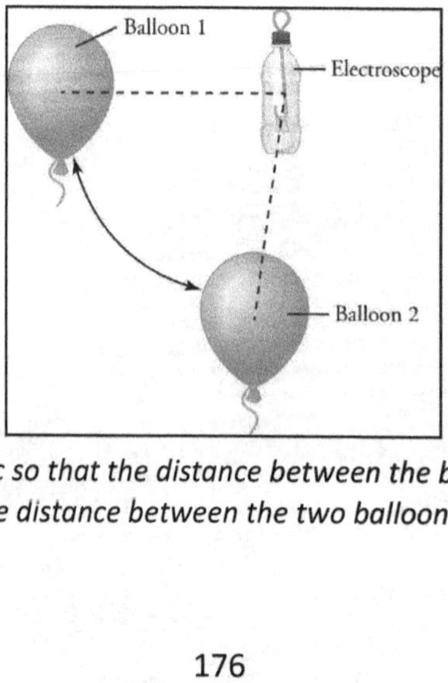

Figure 15.3: *Move one balloon in an arc so that the distance between the balloon and the electroscope is constant while the distance between the two balloons varies.*

Data and Observations

Table 15.2 Electroscopes and Charged Objects

Conditions	Distance Between the Tips of the Foil Leaves (cm)
1 charged balloon	
2 charged balloons, close to each other	
2 charged balloons, far from each other	
1 charged balloon + wool	
1 charged balloon + plastic object	
1 charged balloon + glass rod	
What observations did you make when a wooden rod touched the charged electroscope?	
What observations did you make when your hand touched the charged electroscope?	

Assessments

1. Did rubbing the two balloons against wool give the same or different charges to each balloon? Explain how you know. TEKS (5)(E)
2. Did the wool used to rub the balloons possess a charge? What evidence supports your answer? TEKS (5)(E)
3. Is the charge on the balloon the same as the charge on the plastic comb/rod? How do you know? Is the charge on the balloon the same as the charge on the glass rod? Explain. TEKS (5)(E)
4. The electroscope was charged by induction before you touched it with the wooden rod. Did the wood conduct electricity? Explain your answer. TEKS (5)(E)
5. The electroscope was charged by induction before you touched it with your hand. Did your hand conduct electricity? Explain your answer. TEKS (5)(E)

Activity 3: Capacitors

As we have seen, many objects can acquire static charges. Electrical charges are often more useful, at least for powering devices, when they are flowing. The flow of electric charge is called **current**. In order to do work with electricity, we must have several components: a conducting material—such as wire, a device that converts electrical energy to some other form—such as a light bulb or buzzer, and a source of electric current. In this activity, you will use a capacitor to provide the current and an LED bulb so you can visually detect when the current is flowing.

A **capacitor** is a device designed to store charge temporarily. It consists of two plates that are separated by a small space. When connected to a battery, a positive charge (+Q) builds up on one plate and a negative charge (−Q) builds up on the other plate (Figure 15.4). When the battery is disconnected, these plates remain charged (for a while). We can then create a circuit with the charged capacitor, using the LED to detect the flow of electricity. An example of a capacitor is a heart defibrillator, which is used in hospitals and ambulances, and shown on TV shows, to give a patient an electric shock to get the heart beating again in proper rhythm.

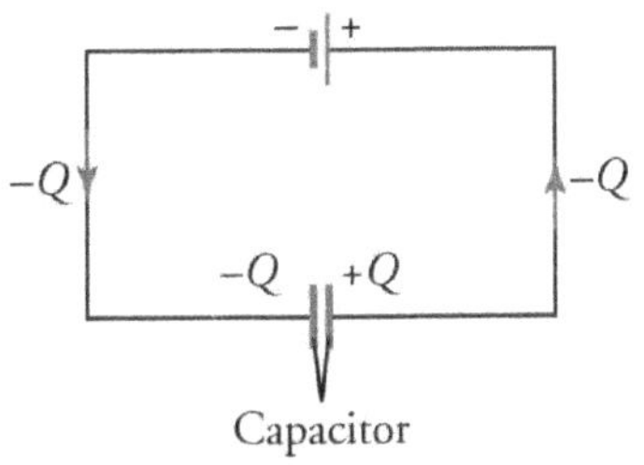

Figure 15.4: *A capacitor can store charge from a battery, with equal and opposite charges on the two plates.*

<u>Safety Precautions</u>

- Follow the directions carefully. Connect the items only as directed; other configurations can result in electric shock.

<u>For this activity you will need the following:</u>

- 9-volt battery
- AAA battery
- Two capacitors with different capacitance ratings (1000 μF and 2000 μF)
- Two LEDs
- Several pieces of insulated wire with alligator clips

******Note***—Test LEDs before using them in a circuit, starting at the lowest voltage, to make sure they do not overload and blow out. A resistor in series prevents overload of LEDs. Each LED needs to be tested for appropriate voltage limit.**

For this activity you will work *in pairs*.

Procedure

Step 1: Connect the capacitor with the lower rating to the 9V battery. Connect the long wire on the capacitor to the negative battery terminal and the short wire to the positive terminal, as shown in Figure 15.5. Keep the capacitor connected for three seconds, then disconnect it from the battery.

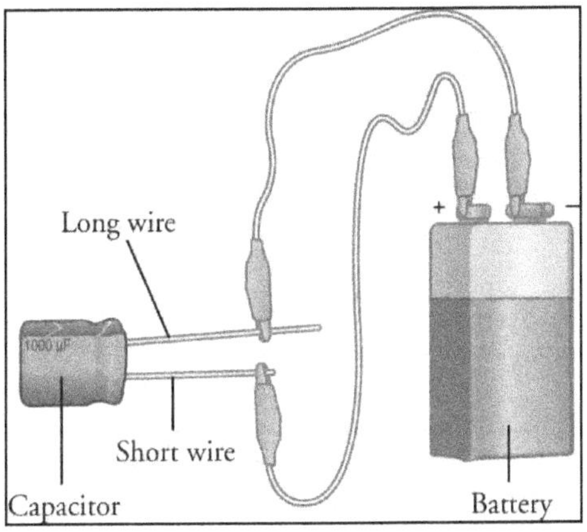

Figure 15.5 *Connecting the capacitor to a battery allows the capacitor to become charged.*

Step 2: Connect one of the capacitor wires to the LED. Note any observations. Connect the other wire to create a closed loop. Write down any observations in Table 15.3, noting how long the light stays on and whether it is bright or dim. *Note*—LEDs have polarity; if the LED does not light up, flip which side the wires are connected to on the L

Step 3: Connect the capacitor to the 9V battery in the opposite orientation. Connect the long wire to the positive terminal and the short wire to the negative terminal. Keep the wires connected for three seconds. Disconnect the battery.

Step 4: Connect the capacitor to the LED, and again, note any observations. Does the orientation of the capacitor matter? Choose one orientation and use it for the remaining steps.

Step 5: Repeat the procedure, but this time charge the capacitor for one second instead of three. Does this alter the time the LED stays lit or alter the brightness?

Step 6: Repeat the procedure again, but this time use the AAA battery instead of the 9V battery. Does this alter the time the LED stays lit or alter the brightness?

Step 7: Repeat the procedure using the 9V battery. This time, connect a wire from the capacitor to one LED. Connect that LED to another LED, and connect the second LED to the capacitor. Again, record all observations in Table 15.3. Does this alter the time the LED stays lit or alter the brightness?

Step 8: Repeat the procedure one last time, but now use the larger capacitor with the 9V battery. Does this alter the time the LED stays lit or alter the brightness?

Data and Observations

Table 15.3 Capacitor Observations

Conditions	Observations
Small capacitor, 9V, 3 sec, long to negative	
Small capacitor, 9V, 3 sec, long to positive	
Small capacitor, 9V, 1 sec	
Small capacitor, AAA, 3 sec	
Small capacitor, 9V, 3 sec, 2 LEDs	
Large capacitor, 9V, 3 sec	

Assessments

1. What conclusions can you draw from the data? Which variables had no effect? For the variables that had an effect, explain what is occurring in terms of current flow. TEKS (5)(F)
2. Sketch the setup you used for Step 7. Trace the current flow through your system. Is this a series circuit (one continuous loop) or parallel (current splits into different pathways)? Design a setup for the other arrangement using the materials available. TEKS (5)(F)

Lab 16: Ohm's Law

Electricity has become vital to much of modern society. Therefore, it is important to understand the basics of simple electrical circuits. Although electrical circuits in houses are very complex, understanding simple circuits can help us communicate with electricians when a problem arises.

Most homes have a combination of series and parallel circuits that complete their wiring. **Series circuits** are wired so that the current must travel through each component in order. **Parallel circuits** are arranged so the components are on different branches, causing the current to split among the components. In this lab, you will use holiday lights to study simple circuits. In your circuits, you will use batteries as the power source. You will determine the resistance of the light bulbs as well as the current and voltage across each of the bulbs. You will observe how voltage and current behave in both series and parallel circuits. Pay attention to differences in the values among light bulbs compared to those of the total circuit.

In this lab you will learn:

- how to measure voltage across resistors in series and parallel circuits;
- how to measure current through resistors in series and parallel circuits;
- how to measure the total resistance in series and parallel circuits;
- how to determine the current in series and parallel circuits.

Activity 1: Voltage TEKS (5)(F)

In this activity, you will measure the voltage drop across each of the light bulbs in both series and parallel circuits. **Voltage** is the electric potential difference between two points on a circuit, where electric potential is the potential energy per unit charge. Thus, if the electric potential difference is 6 volts, then one coulomb of charge will gain 6 joules of potential energy after moving from the first point to the second. To measure the voltage of a resistor with a multimeter, the meter must be placed parallel to the resistor.

Safety Precautions

- To help prevent electrocution, do not leave water around electrical equipment and sockets.
- Inspect electrical cords for any exposed wires to prevent electrical shock.
- Use caution when touching light bulbs, which may be hot after use.

For this activity you will need the following:

- Holiday light bulbs
- Multimeter tool set to appropriate voltage for the batteries used
- Battery and mount
- Switch
- Wires
- Alligator clips (optional)

For this activity you will work in *pairs or groups*.

Procedure

Step 1: Pay close attention to your teacher, who will demonstrate how to attach all of the components of the circuits you will be constructing in this activity. Then, make a series circuit like that shown in Figure 16.1 with a battery and three lights, using the necessary wires and alligator clips. Record the voltage of the battery in Table 16.1.

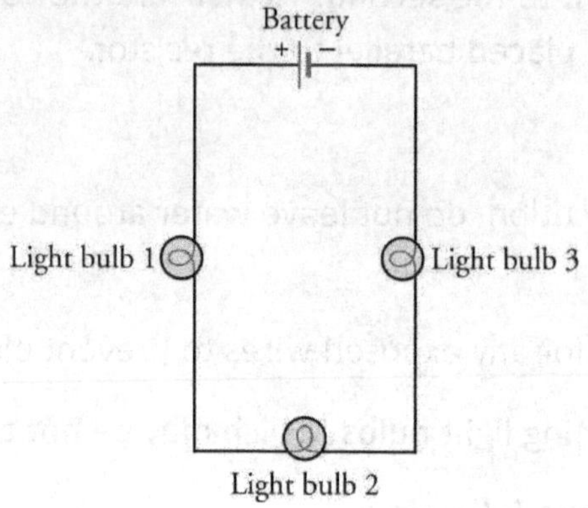

Figure 16.1: *Connect your lights and battery as in the diagram to create a series circuit.*

Step 2: Pay close attention to your teacher, who will now demonstrate how to connect the multimeter to your circuit and measure voltage. Then, connect the multimeter as shown in Figure 16.2 so that the meter is in parallel with the first light. Record the voltage drop across the light in Table 16.1.

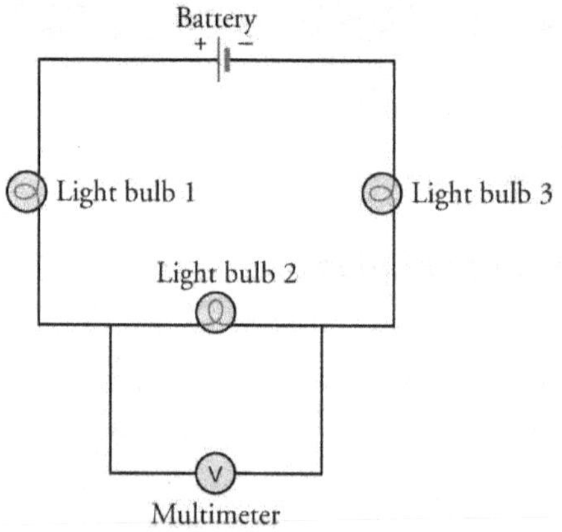

Figure 16.2: *Connect the multimeter parallel to a light to measure the voltage drop across the light bulb.*

Step 3: Using the same procedure, measure the voltage drop across light 2, then light 3. Record your readings from the multimeter in Table 16.1.

Step 4: Record the voltage drop across lights 1 and 2 by connecting the multimeter parallel with both lights. Record your readings from the multimeter in Table 16.1.

Step 5: Record the voltage drop across lights 2 and 3 by connecting the multimeter parallel with both lights. Record your readings from the multimeter in Table 16.1.

Step 6: Make a parallel circuit with a battery and three lights, using the necessary wires and alligator clips, as shown in Figure 16.3.

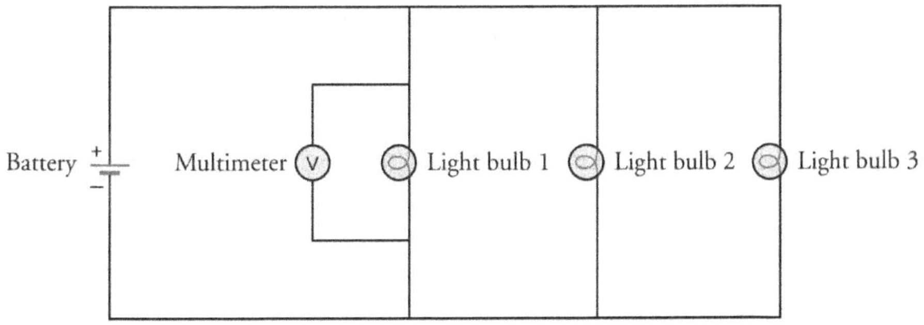

Figure 16.3: *Connect your lights, battery, and multimeter as in the diagram to create a parallel circuit.*

Step 7: Determine the voltage across each of the light bulbs in the parallel circuit. Connect the multimeter as shown in the diagram to each of the light bulbs to measure the voltage. Record your results in Table 16.1. How do the voltages compare across the lights in a series circuit versus a parallel circuit? How do the voltages of the individual light bulbs compare to the combined voltages in the series circuit?

Data and Observations

Table 16.1: Voltage Across Light Bulbs in Series and Parallel Circuits

	Series Circuit	Parallel Circuit
Battery Voltage		
Light 1		
Light 2		
Light 3		
Lights 1 and 2		
Lights 2 and 3		

Assessments

1. A circuit contains three resistors that have a voltage drop of 6 volts each. The circuit contains a 6-volt battery. Are the resistors connected in series or parallel? Explain how you know. TEKS (5)(F)
2. A series circuit contains three resistors and has voltage drops of 3 volts across the first resistor, 3 volts across the second resistor, and 6 volts across the third resistor. What is the total voltage available in the circuit? TEKS (5)(F)
3. How does the voltage drop of each light compare to the battery:
   a. in a series circuit?
   b. in a parallel circuit?

   TEKS (5)(F)

Activity 2: Current TEKS (5)(F)

In this activity, you will measure the total current in the circuit as well as across each resistor. **Current** is the rate of flow of electric charge through the circuit. In a series circuit, the current across each resistor is equal to the total current in the circuit. In a parallel circuit, the current is divided among the resistors based on their resistance. The multimeter must be placed in series with a resistor to measure current.

Safety Precautions

- To help prevent electrocution, do not leave water around electrical equipment and sockets.
- Inspect electrical cords for any exposed wires to prevent electrical shock.
- Use caution when touching light bulbs, since they may be hot after use.

For this activity you will need the following:

- Holiday light bulbs
- Multimeter tool, set to appropriate voltage for the batteries used
- Battery and mount
- Switch
- Wires
- Alligator clips (optional)

For this activity you will work in *pairs or groups*.

Procedure

Step 1: Pay close attention to your teacher, who will demonstrate how to attach all of the components of the circuits you will be constructing in this activity. Make a series circuit with a battery and three lights, using the necessary wires or alligator clips.

Step 2: Pay close attention to your teacher, who will now demonstrate how to connect the multimeter to your circuit and measure its current. Then, connect the multimeter as shown in Figure 16.4 so that the meter is in series with the battery. Measure the current traveling through the circuit and record the reading in Table 16.2.

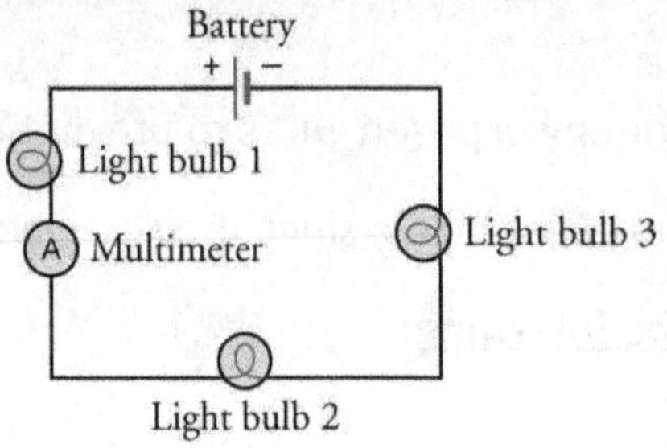

Figure 16.4: *Connect the multimeter in series with a light bulb to measure the current through it.*

Step 3: Make a parallel circuit with a battery and three lights, using the necessary wires or alligator clips.

Step 4: Connect the multimeter in series with all the lights to measure the current traveling through the circuit. Record the reading on the multimeter in Table 16.2.

Step 5: Connect the multimeter in series with light 1 to measure current traveling through the light. Record the reading on the multimeter in Table 16.2. Repeat readings with lights 2 and 3.

Step 6: Connect the multimeter as in Figure 16.5 in series with lights 2 and 3 to measure current traveling through both lights. Record the reading on the multimeter in Table 16.2.

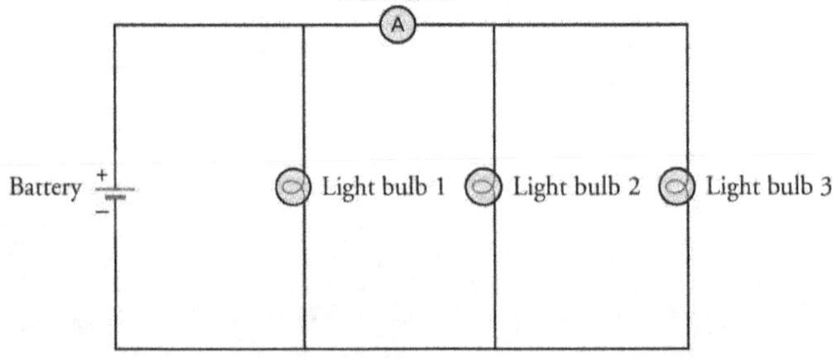

Figure 16.5: *Connect the multimeter in series with light bulbs 2 and 3 to measure the current through both bulbs.*

Data and Observations

Table 16.2: Current in Series and Parallel Circuits

	Series Circuit	Parallel Circuit
Total current in circuit		
Current through light 1	Same as total	
Current through light 2	Same as total	
Current through light 3	Same as total	
Current through lights 1 and 2	Same as total	
Current through lights 2 and 3	Same as total	

Assessments

1. A series circuit has three resistors of 5 ohms each. What is the current through the circuit if the current through each resistor is measured at 3 amps? TEKS (5)(F)
2. A parallel circuit has three resistors and the current traveling through each resistor is 5 amps. What is the total current in the circuit? TEKS (5)(F)
3. A parallel circuit made of these light bulbs has a total current of 5 amps. Two of the light bulbs measured at 2 amps each. What is the current flowing through the third light bulb? Explain how you know. TEKS (5)(F)

Activity 3: Resistance TEKS (5)(F)

In this activity, you will calculate the resistance of the light bulbs and the total resistance of the circuits. **Resistance** is a measure of how much a material reduces the flow of electric charge. To find the resistance, you will first need to measure the current and voltage of each light bulb and the total circuit.

Safety Precautions

- To help prevent electrocution, do not leave water around electrical equipment and sockets.
- Inspect electrical cords for any exposed wires to prevent electrical shock.
- Use caution when touching light bulbs, since they may be hot after use.

For this activity you will need the following:

- Holiday light bulbs
- Multimeter tool, set to appropriate voltage for the batteries used
- Battery and mount
- Switch
- Wires
- Alligator clips (optional)

For this activity you will work in *pairs or groups.*

Procedure

Step 1: Make a series circuit with a battery and three lights, using the necessary wires or alligator clips.

Step 2: Measure the current and voltage across each resistor using a multimeter. Record the readings in Table 16.3.

Step 3: Use Ohm's law, $I = \frac{\Delta V}{R}$, to solve for the resistance of each light bulb. How does the resistance of the individual light bulbs compare to the total resistance of the series circuit?

Step 4: Make a parallel circuit with a battery and three lights, using the necessary wires or alligator clips.

Step 5: Measure the current and voltage across each resistor using a multimeter. Record the readings in Table 16.4.

Step 6: Use Ohm's law, $I = \frac{\Delta V}{R}$, to solve for the resistance of each light bulb. How does the resistance of the individual light bulbs compare to the total resistance of the parallel circuit? How does the total resistance of the series circuit compare to the total resistance of the parallel circuit when they contain the same number of light bulbs?

Data and Observations

Table 16.3: Readings for Series Circuit

	Current	Voltage	Resistance
Light 1			
Light 2			
Light 3			
Total Circuit			

Table 16.4: Readings for Parallel Circuit

	Current	Voltage	Resistance
Light 1			
Light 2			
Light 3			
Total Circuit			

Assessments

1. A series circuit has three light bulbs of equal resistance. What happens to the total resistance of the circuit if a light bulb is removed from the circuit? [TEKS 5F]
2. A parallel circuit has four light bulbs of equal resistances. What happens to the total resistance of the circuit if an additional light bulb is added in parallel to the others? TEKS (5)(F)
3. Explain what happens to the current traveling through a light bulb in a parallel circuit when an additional light bulb of equal resistance is added in parallel. TEKS (5)(F)

Activity 4: Ohm's Law TEKS (5)(F)

In this activity, you will use the multimeter to determine the actual resistance of the light bulbs. The multimeter will send a low current through the light bulb and measure the voltage drop to determine the resistance. You can use this known resistance, the voltage of the battery, and Ohm's law to calculate current through each bulb and/or the circuit. At the end of the activity, you can check your calculation by testing the current of the circuit with the multimeter. Ohm's law is represented by the equation

$$I = \frac{\Delta V}{R},$$

where, *I* is current measured in amperes, ΔV is voltage drop measured in volts, and *R* is resistance measured in ohms.

Safety Precautions

- To help prevent electrocution, do not leave water around electrical equipment and sockets.
- Inspect electrical cords for any exposed wires to prevent electrical shock.
- Use caution when touching light bulbs, since they may be hot after use.

For this activity you will need the following:

- Holiday light bulbs
- Multimeter tool
- Battery and mount
- Switch
- Wires
- Alligator clips (optional)

For this activity you will work in *pairs or groups*.

Procedure

Step 1: Measure the resistance of three light bulbs using the multimeter. Plug the red cord into the **VΩmA port and the black plug into the COM port**. Place the black and red on either end of the light bulb. Start with the highest-value setting; if the reading is close to zero, change the setting to a smaller value until you measure an accurate reading. Record resistances in Table 16.5.

Step 2: Make a series circuit with a battery, three lights, and the necessary wires or alligator clips.

Step 3: From the measured resistance of the light bulbs and the voltage of the battery, calculate the current for the circuit using Ohm's law. Show your work and record the calculated current in Table 16.5.

Step 4: Use the multimeter to measure the total current in the circuit. Record the reading in Table 16.5. Compare the values.

Step 5: Repeat the procedure with a parallel circuit. Record data in Table 16.6.

Data and Observations

Table 16.5: Series Circuits and Ohm's Law

Series Circuit	Light Bulb 1	Light Bulb 2	Light Bulb 3
Resistance of light bulbs			
Total resistance			
Voltage of battery			
Calculated current of circuit			
Measured current			

Table 16.6: Parallel Circuits and Ohm's Law

Parallel Circuit	Light Bulb 1	Light Bulb 2	Light Bulb 3
Resistance of light bulbs			
Total resistance			
Voltage of battery			
Calculated current of circuit			
Measured current			

Assessments

1. What is the voltage needed to produce a current of 3 A for three resistors of 6 ohms each that are connected in parallel? TEKS (5)(F)
2. Using Ohm's law, explain what happens to the current in a circuit when the resistance is increased. TEKS (5)(F)
3. What is the current of a circuit that has two resistors, of 6 ohms and 3 ohms, connected in series with a 12-volt battery? TEKS (5)(F)

Lab 17: Circuits

An **electric circuit** is defined as electric components connected in a path through which electricity can flow. Electrons move along this path from an area of higher electric potential to an area of lower electric potential creating an **electric current** running in the direction opposite the electron movement (Figure 17.1). Electric current measures the number of electrons or other charge carriers that flow past a certain point per second, and its SI unit is the **ampere**—or **amp**. An electric current that flows in only one direction is **direct current** (DC). A current where flow can change directions is **alternating current** (AC). Batteries are DC, whereas the current flowing through your house is AC. Phone chargers are an example of a **transformer,** which converts the current and allows you to charge a DC phone with an AC current.

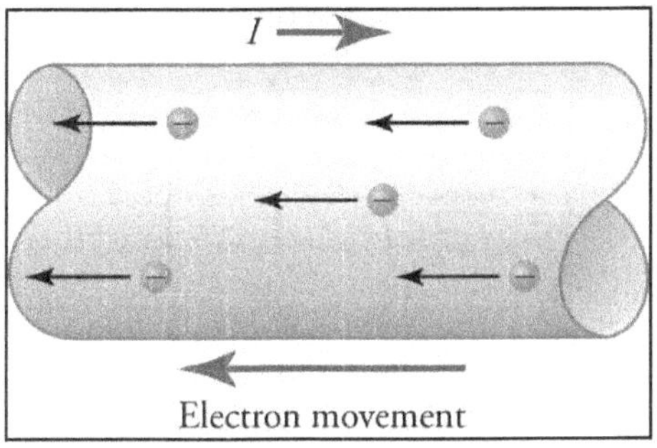

Figure 17.1: *Negatively charged electrons move in one direction and positive current (I) flows in the opposite direction.*

Circuits have several components, which can be set up in a series connection, parallel connection, or some combination of both. Figure 17.2 shows a simple circuit with just a battery and a lightbulb. The battery uses chemical reactions to produce a difference in **electrical potential** between its positive and negative terminals. The electrons move from the negative terminal of the battery to the positive along the electrical circuit resulting in the flow of electricity in the opposite direction—that is, from the positive to the negative terminal. The battery generates the **electromotive force (emf)** to push the electrons along the wire and create the current. This movement is from the difference in electric potential energy. The **emf** represents energy per unit charge (voltage) and is measured in **volts**.

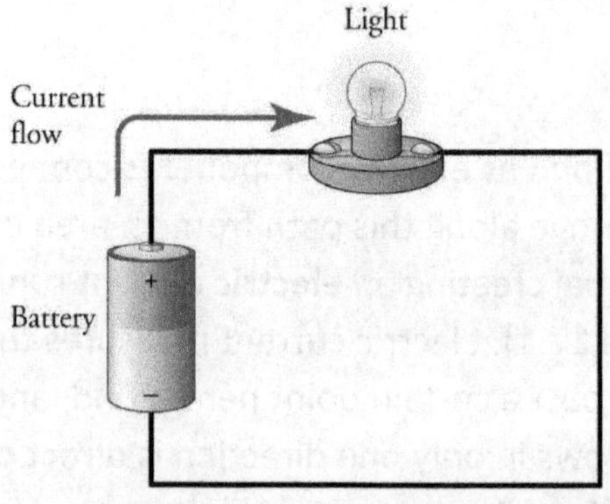

Figure 17.2: *A simple circuit consisting of a source of a battery and a light, which acts as a resistor.*

In a **series circuit**, each component is lined up one after another. In a **parallel circuit**, the components are lined up next to each other. A **complex circuit** is any combinations of series and parallel circuits (Figure 17.3).

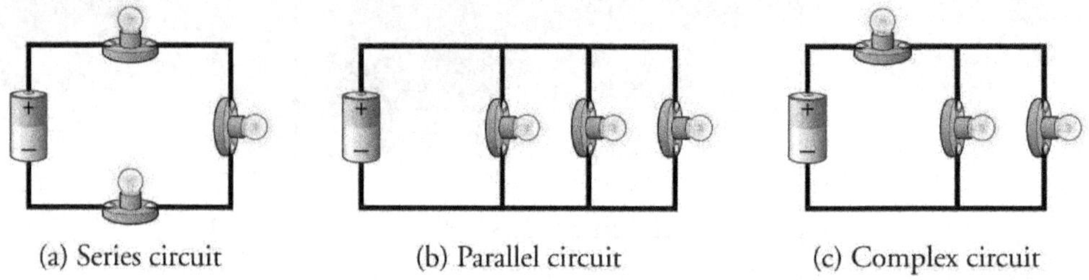

Figure 17.3: *Comparison of (a) series, (b) parallel, and (c) complex circuits. Resistors are represented by a zig-zag line.*

Resistors are devices that resist, or slow down, an electric current and are represented as a zig-zag line on a circuit diagram. Resistors have a measurable potential difference across them, which affects the movement of electrons as they flow through the circuit. The capacity of a resistor to impede an electric current is called its **resistance** and is measured in **ohms** (Ω), defined as volts per amp. The initial voltage—generated by the battery in this lab—is equal to the voltage drop across the resistor—or sum of the voltage drops across all resistors in a circuit. **Ohm's law** defines the relationships between current, voltage, and resistance. Figure 17.4 shows a simple way the three variables relate to one another and can be used to determine one variable when the other two are known. Current is directly proportional to applied voltage and inversely proportional to resistance. Ohm's law tells us that as resistance increases, the current flowing through the resistor decreases. An **ohmic material** is one that behaves according to Ohm's law and is ideal for use as a resistor.

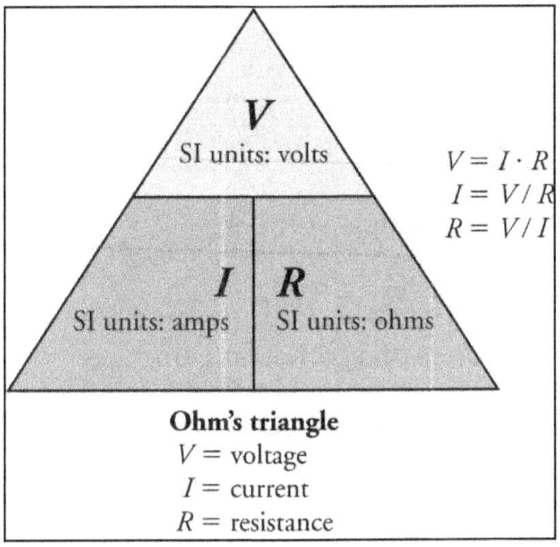

Figure 17.4: Ohm's triangle: *How voltage, current and resistance are related in Ohm's law.*

The filament inside an incandescent lightbulb is a resistor: as the electrons move through the filament the energy in the form of heat is transferred to the filament, causing it to glow. The **power** (SI units: **watts**) dissipated through the filament determines the bulb's brightness: As the power increases, so does the brightness. The equations for power are

$$P = I^2 \cdot R \quad \text{or} \quad P = \frac{V^2}{R},$$

where *P* is power, measured in watts; *I* is current, measured in amperes; and *R* is resistance, measured in ohms.

As you complete this lab, use the symbols in Figure 17.5 when drawing or reading circuit diagrams.

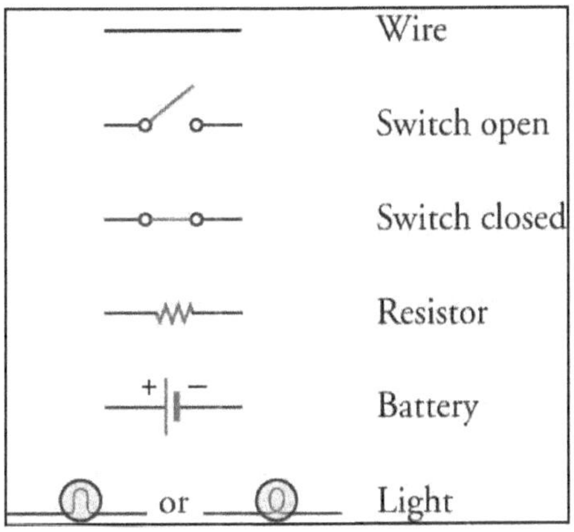

Figure 17.5: *Symbols used in circuit diagrams.*

In this lesson you will learn:

- how to construct a series circuit;
- how to construct a parallel circuit;
- how to design and construct a complex circuit;
- how to calculate the resistance in a circuit;
- how to calculate the resistance of a particular component of a circuit;
- how to predict/explain the behavior of lightbulbs in a given circuit.

Activity 1: Series Circuit

In this activity, you will construct a circuit with three identical light bulbs arranged in series and observe the behavior of the bulbs as they are removed from the circuit one at a time. Recall that a series circuit has its components arranged end to end (see Figure 17.3 (a)). A series circuit can be thought of as the cars of a train, lined up one directly behind another.

When components are arranged in series, the equivalent resistance is the sum of the component resistances, and is measured in ohms (Ω). The equivalent resistance (R_{eq}) for *n* resistors is

$$R_{\text{eq}} = \frac{V}{I},$$

or

$$R_{eq} = \sum R = R_1 + R_2 + \ldots + R_n .$$

The current running through all resistors in a series is the same for each resistor. The power dissipated by a resistor is equal to current times voltage, with SI units of watts (W).

$$P = I \cdot V$$

Safety Precautions

- You will be working with electricity during this lab—be mindful of what you are doing.
- Do not add any additional components to the circuits.
- Ask your instructor for help if you are unsure of something.
- Notify your instructor immediately in the event of injury.
- Keep liquids away from circuits.

For this activity you will need the following:

- Battery of known voltage
- Wire
- Three identical light bulbs with sockets
- Masking tape or labels

For this activity you will work *in pairs or in groups*.

Procedure

Step 1: Using Figure 17.3 as a reference, attach the battery to the three lightbulbs using the wire—leave bulbs out of sockets during this step. Using masking tape, label the light closest to the battery as 'light 1' and label the subsequent bulbs '2' and '3'.

Step 2: Insert a lightbulb to each socket. How bright are the lights? Are they the same brightness as each other or are the brightness levels different? Record the brightness of the bulbs—relative to one another—in Table 17.1.

Step 3: Remove the third bulb from the socket. Describe in Table 17.1 what happens to the brightness level in the other two bulbs.

Step 4: Now, remove the second bulb from the socket and describe in Table 17.1 the relative brightness for each bulb. What do you think is the relationship between resistance in a series circuit and how it affects the power output, or brightness, of the bulbs in series? Discuss with other students to see if you arrived at the same conclusion.

Data and Observations

Table 17.1: Bulb Brightness in a Series Circuit

No. of Bulbs	Observations
3	
2	
1	

Assessments

1. If the current running through the circuit is 0.0420 A and the voltage of the battery is 24V, what is the resistance of each of the bulbs in your circuit? Assume that each bulb is an ohmic resistor. TEKS (5)(F)
2. The potential difference across an element in a series circuit can be calculated using Ohm's law: potential difference = voltage. Calculate the potential difference across each of three different resistors (85 Ω, 140 Ω, and 250 Ω) in a series circuit with a current of 0.0350 A. TEKS (5)(F)
3. Given Figure 17.6, calculate the resistance of each resistor represented by the two lines. TEKS (5)(F)

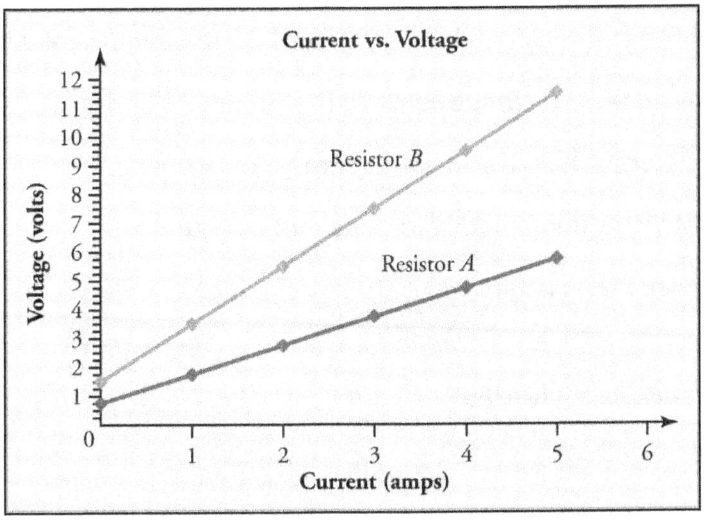

Figure 17.6: *Graph of current versus voltage (Assessment 3).*

Activity 2: Parallel Circuit

Think of a parallel circuit as a road going from one lane to three. Instead of the current flowing along only one path, it has multiple *options* for movement. In a parallel circuit, the voltage drop across each resistor is the same but the current running through each resistor may be different if the resistors have different resistances. For a parallel arrangement, equivalent resistance is calculated using this equation—given *n* resistors

$$\frac{1}{R_{eq}} = \sum \frac{1}{R} = \frac{1}{R_1} + \frac{1}{R_2} + ... + \frac{1}{R_n}.$$

Safety Precautions

- You will be working with electricity during this lab—be mindful of what you are doing.
- Do not add any additional components to the circuits.
- Ask your instructor for help if you are unsure of something.
- Notify your instructor immediately in the event of injury.
- Keep liquids away from circuits.

For this lab you will need the following:

- Battery of known voltage
- Wire
- Three identical light bulbs with sockets
- Masking tape or labels.

For this activity you will work *in pairs or groups*.

Procedure

Step 1: Build a parallel circuit as shown in Figure 17.7—leave the sockets empty. Using either masking tape or labels, label the sockets as shown in the diagram.

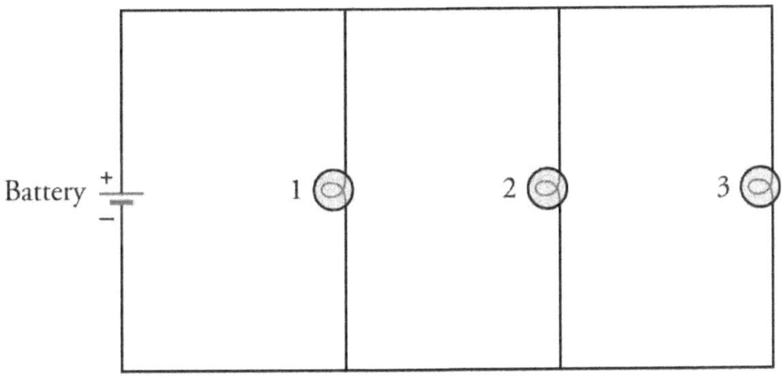

Figure 17.7: *Activity 2: Parallel circuit layout.*

Step 2: Add the lightbulbs to each socket. How does the brightness of each bulb compare to one another? Record your observations in Table 17.2.

Step 3: One by one, remove bulbs from sockets. Observe and record brightness of remaining bulb/bulbs in the data table. How does resistance in a parallel circuit affect bulb brightness? Is there any difference in the brightness of bulb one versus bulb three when all bulbs are in the sockets? Why or why not?

Data and Observations

Table 17.2: Activity 2: Bulb Brightness in a Parallel Circuit

Action	Result
All three bulbs	
Remove bulb 1	2 =
	3 =
Remove bulb 2	1 =
	3 =
Remove bulb 3	1 =
	2 =
Remove bulbs 1 and 2	3 =
Remove bulbs 1 and 3	2 =
Remove bulbs 2 and 3	1 =

Assessments

1. Find the total current through a parallel circuit with three resistors ($R_1 = 100\,\Omega$, $R_2 = 75\,\Omega$, and $R_3 = 125\,\Omega$), and a 12.0-V battery. Would this be the same if the resistors were in series? [TEK 5.F]

    a. Find the current through each of the three resistors in part a. Based on your answer, speculate what happens to the current as resistance increases. TEKS (5)(F)

    b. How would this be different if the resistors were in series?

2. Design a circuit that has two sets of parallel resistors that are parallel to one another. TEKS (5)(F)

3. Calculate the power of each resistor given in question 1. TEKS (5)(F)

Activity 3: Complex Circuits

A **complex circuit** consists of any combination of series and parallel arrangements. Consider a single-lane highway with toll booths every 30 miles. After a while, the highway goes from one to three lanes, each with a tollbooth where the lanes begin. The single-lane road with tolls represents a series, and the three lanes with their three tolls represent a parallel arrangement. Any complex circuit can be reduced to a single equivalent resistance by systematically simplifying the structure of the circuit as shown in Figure 17.9. Using the equations for equivalent resistance, combine the two resistors that are in series into one and calculate the new R_{eq}; then combine the result with the last resistor (parallel) and calculate the final R_{eq} —in this example all the resistors have the same resistance.

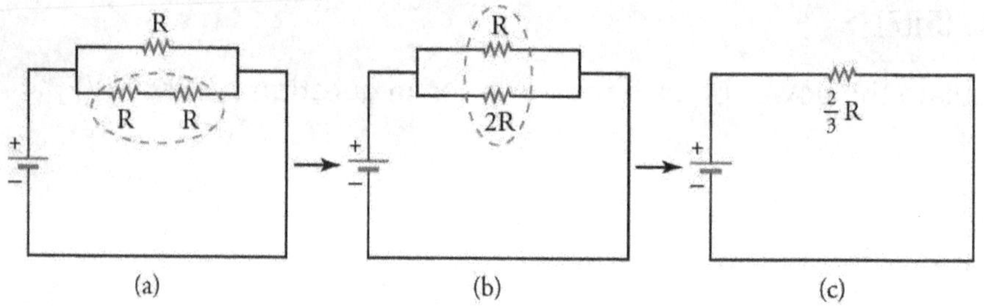

Figure 17.9: *(a) Original complex circuit; (b) combine two resistors in series to create one resistor; (c) combine two parallel resistors into one.*

<u>Safety Precautions</u>

- You will be working with electricity during this lab—be mindful of what you are doing.
- Do not add any additional components to the circuits.
- Ask your instructor for help if you are unsure of something.
- Notify your instructor immediately in the event of an injury.
- Keep liquids away from circuits.

<u>For this lab you will need the following</u>:

- Battery of known voltage
- Wire
- Three identical light bulbs with sockets
- Masking tape or labels

You will be performing this activity *in pairs or groups.*

Procedure

Step 1: Build a complex circuit using two bulbs in parallel and one in series. Use either the masking tape or labels to label each light, as shown in Figure 17.10 (a)).

Step 2: Remove one lightbulb at a time and record your observations in Table 17.3. Be sure to identify which bulb/bulbs were removed using numbers—refer to the diagram—and the effects on each remaining bulb/bulbs. For example, indicate if bulbs one and two are brighter and the same, brighter but different—indicate which is brighter—etc.

Step 3: Build a complex circuit using two lights in series with another bulb parallel to them, as in Figure 17.10 (b)).

Step 4: Remove lightbulbs as listed in Table 17.4 and record your observations.

In the activity for complex circuits, bulbs of the same wattage are used. Would the results you observed be the same or different if each bulb was a different wattage? Assume bulb one is the lowest wattage and bulb three is the highest.

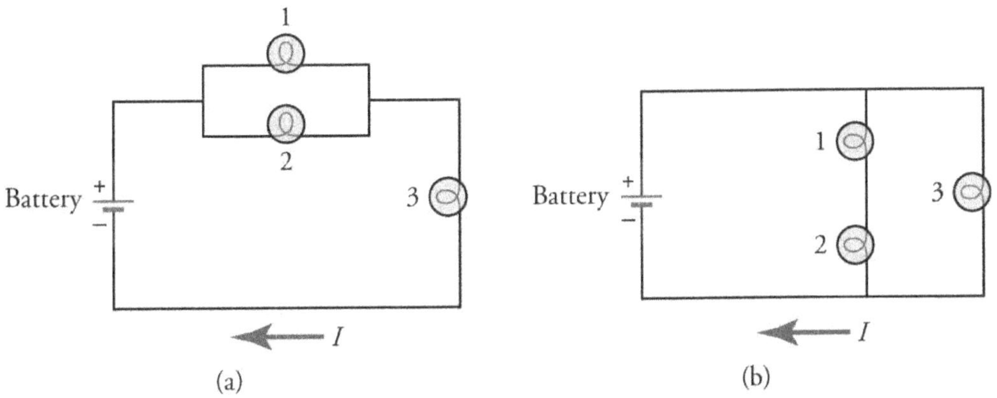

Figure 17.10: *(a) Complex circuit of two parallel bulbs in series with a third. (b) Complex circuit consisting of two series bulbs parallel to a third bulb.*

Data and Observations

Table 17.03: Bulb Brightness in Complex Circuits

Action	Result
All three bulbs	
Remove bulb 1	2 =
	3 =
Remove bulb 2	1 =
	3 =
Remove bulb 3	1 =
	2 =

Table 17.4: Complex Circuits (Step 2)

Action	Result
All three bulbs	
Remove bulb 1	2 =
	3 =
Remove bulb 2	1 =
	3 =
Remove bulb 3	1 =
	2 =

Assessments

1. Calculate the equivalent resistance in the following circuit. TEKS (5)(F)

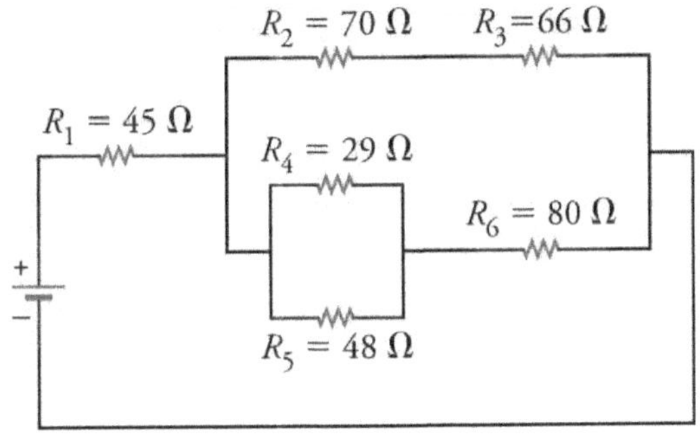

2. Give an example of a circuit that has five resistors. The resistors should have resistance R that can be reduced down to an equivalent resistance of $2R$. TEKS (5)(F)

3. Based upon your observations, decide if string lights—holiday lights—are in series, parallel, or some combination of both. Support your answer. TEKS (5)(F)

Lab 18: Work, Energy, and Power in Circuits

Our modern world runs on electrical power. Inside every building, automobile, appliance, computer, and hand-held electronic device are circuits that deliver electricity to each electrical component. Because so much of our lives depend on electrical devices, applying knowledge of electrical power is an important skill. Knowledge of electrical power can help you troubleshoot a broken device, shop for battery systems, save money and electricity, or even build your own electrical device or power system.

Most circuits in modern electrical devices are complex. Yet, most complex circuits are combinations of series and parallel circuits. In this lab, we will begin examining electrical power in series and parallel circuits, and then move on to determining power in a complex circuit. You will also learn how to determine the power of a component in a circuit—in this case, a light bulb. You will determine the overall power in a circuit for both parallel and complex circuits. This will require measuring current and voltage in circuits and using these values to calculate power.

In this lab you will learn:

- how to determine the power of a light bulb in a series circuit;
- how to determine the power of a parallel circuit;
- how to determine the power of a complex circuit.

Activity 1: Power of a Series Circuit TEKS (6)(C)

What is **power** in the context of an electrical circuit? Recall, that power is the rate at which energy is used. Electric power measures the rate at which electrical energy is converted to other forms of energy (light, heat) in the circuit. The formula to use for calculating electric power is

$$\text{power} = \text{voltage} \times \text{current}$$
$$P = VI$$

The SI unit of power is the watt (W). You may have noticed that household bulbs are usually sold in packages that display their wattage. This gives consumers information about how much power the bulb will consume.

In this activity, you will measure the voltage and current in a circuit with several holiday lights in series. You will use this information to calculate the power of the series circuit and of one bulb in the circuit.

Safety Precautions

- Handle glass bulbs carefully.
- Do not connect both meter leads directly to the battery when measuring current.
- Check with your teacher to provide specific instruction for the meters you use during this activity.

For this activity you will need the following:

- Seven holiday lights cut from an incandescent 100-light strand*
- 9V battery
- Connector wires
- Wire cutter and stripper
- Multimeter (or separate current and voltage meters)

*Use a 100-bulb strand of mini lights. Be aware that some 100-bulb strands are actually two 50-bulb strands connected together. If this is the case, a 9V battery will only light a strand of up to three bulbs. You may use a shorter strand or a 12V battery pack (e.g., from an old phone charger).

For this activity you will work *in pairs*.

Procedure

Step 1: Connect the strand of holiday lights to the 9V battery as shown in Figure 18.1. Use the wire stripper to strip the ends of the wires as necessary (ask your teacher for a demonstration if you are not familiar with this tool). This arrangement forms a simple series circuit.

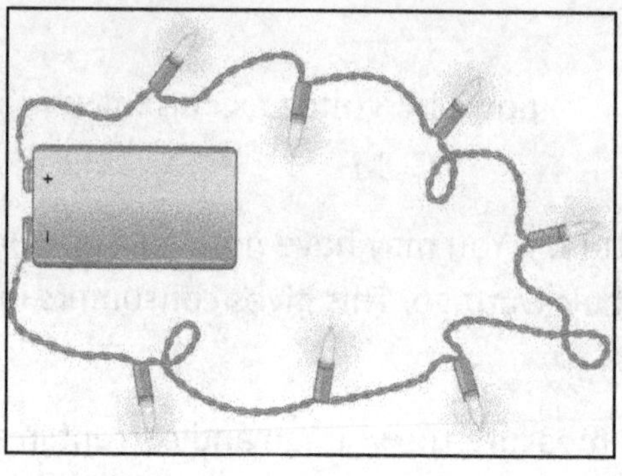

Figure 18.1: *Connect each end of the strand to one of the battery terminals.*

Step 2: Use your meter to measure the voltage drop in your circuit. This measurement represents the loss of voltage as electricity moves through the circuit. For most meters, you will connect the leads to the wires where they meet the battery as shown in Figure 18.2. Check with your teacher for specific instructions on how to set up your meter to get the correct voltage reading. It should be slightly less than 9V. Enter the voltage in Table 18.1.

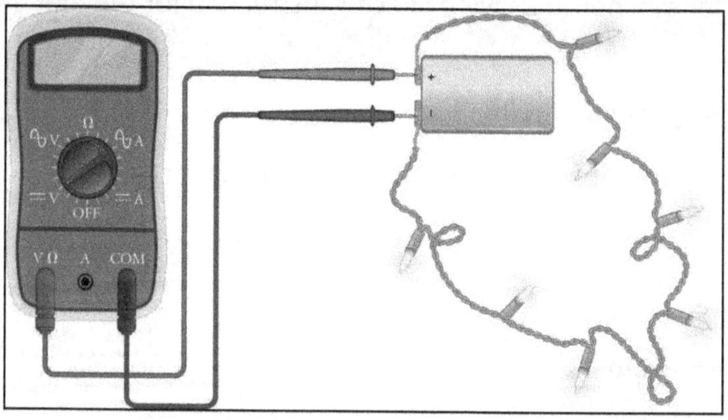

Figure 18.2: *Touch each lead of the multimeter to the spots where the wires meet the battery terminals.*

Step 3: Use your meter to measure the current in your circuit. *Caution—Do not connect both meter leads directly to the battery, since this can damage the meter.* You will need to disconnect one of the ends of the light strand and place your meter in series with the battery and the lights (Figure 18.3). Check with your teacher for specific instructions on how to set up your meter to get the correct current reading. Enter the current in Table 18.1.

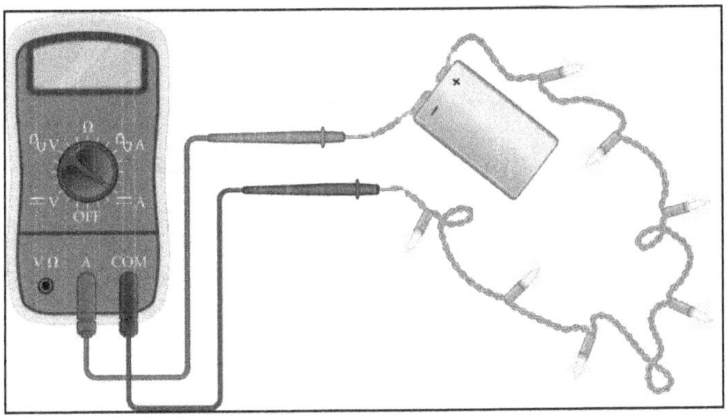

***Figure 18.3**: Place the leads of the multimeter in series with the light string and the battery.*

Data and Observations

<u>Table 18.1 Voltage and Current in the Series Circuit</u>

	Voltage Drop (V)	Current (A)
Series circuit		

Assessments

1. Use your measurements of voltage and current to calculate the power in the series circuit. TEKS (6)(C)
2. Assume that each bulb in the circuit has an equal resistance value. TEKS (6)(C)
   a. Calculate the voltage drop across one bulb. Explain your response. TEKS (6)(C)
   b. Calculate the power of one bulb. TEKS (6)(C)
3. Explain how you can use the equation for electric power to calculate energy use over time. TEKS (6)(C)

Activity 2: Power of a Parallel Circuit TEKS (6)(C)

In Activity 1, you determined the power of a series circuit. What happens if you place bulbs in parallel instead of series?

Safety Precautions

- Handle glass bulbs carefully.
- Do not connect both meter leads directly to the battery when measuring current.
- Check with your teacher to provide specific instruction for the meters you will use for this activity.

For this activity you will need the following:

- 4 holiday lights cut from the strand you used in Activity 1
- 9V battery
- connector wires
- wire cutter and stripper
- multimeter (or separate current and voltage meters)

For this activity you will work *in pairs*.

Procedure

Step 1: Cut four lights from the strand of holiday lights. Position your cuts equidistant between neighboring lights to be sure you have enough wire attached to each light. Strip the wire at each end of your resulting strand so that 1 cm of wire is exposed. Then, connect the four lights and additional connector wires to create a parallel circuit (Figure 18.4).

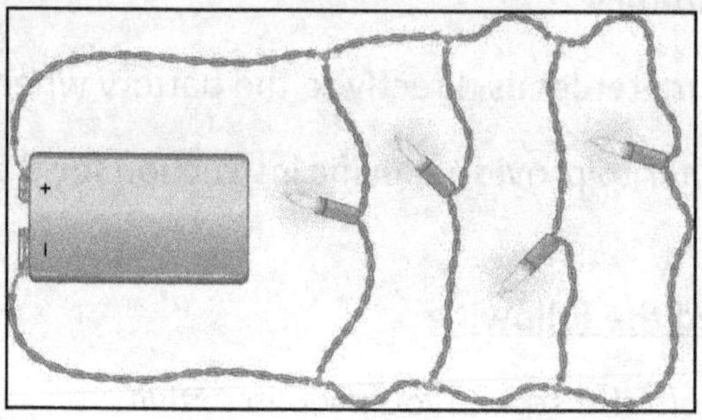

Figure 18.4*: Connect the four lights with connector wires to create a parallel circuit.*

Step 2: Use the meter to measure the voltage drop across the battery as you did in Activity 1. Then, measure the voltage drop across each bulb. For most meters, you will touch the wire with the probes on each side of the battery. Check with your teacher for specific instructions on how to set up your meter to get the correct voltage reading. Enter the voltages in Table 18.2.

Step 3: Use the meter to measure the current in each strand of the parallel circuit. You will need to disconnect one of the ends of the light strand and place the meter in series with the light and the connector wire. *Caution—Do not connect both meter leads directly to the battery, since this can damage the meter.* Check with your teacher for specific instructions on how to set up your meter to get the correct current reading. Enter the current in Table 18.2.

Data and Observations

Table 18.2 Voltage and Current in the Parallel Circuit

	Voltage Drop (V)	Current (A)
Across the battery (Voltage Drop only)		
Across each parallel strand		

Assessments

1. Compare how voltage and current are distributed in the bulbs in your parallel circuit. TEKS (6)(C)
2. Calculate the total current in the circuit. TEKS (6)(C)
3. Calculate the power of the parallel circuit. TEKS (6)(C)
4. Compare the power consumption in the series and parallel circuits. What are the effects of this difference in power consumption, in terms of bulb brightness and battery consumption? TEKS (6)(C)

Activity 3: Power of a Complex Circuit TEKS (6)(C)

Most devices rely on complex circuits. How can you determine the power in a complex circuit? Most complex circuits are simply combinations of parallel and series circuits. In this activity, you will put together what you have learned about determining power in series and parallel circuits to determine the power in a complex circuit.

Safety Precautions

- Handle glass bulbs carefully.
- Do not connect both meter leads directly to the battery when measuring current.
- Check with your teacher to provide specific instruction for the meters you will use for this activity.

For this activity you will need the following:

- All of the pieces of the strand of holiday lights used in Activity 1:
    - A strand of three holiday lights
    - Four individual holiday lights
- 9V battery
- Connector wires
- Wire cutter and stripper
- Multimeter (or separate current and voltage meters)

For this activity you will work *in pairs.*

Procedure

Step 1: How can you make a complex circuit with the parts you have left from Activity 1 and Activity 2? Your complex circuit must have both parallel and series components. Design a complex circuit such as the one in Figure 18.5. Then submit your design to your teacher to approve.

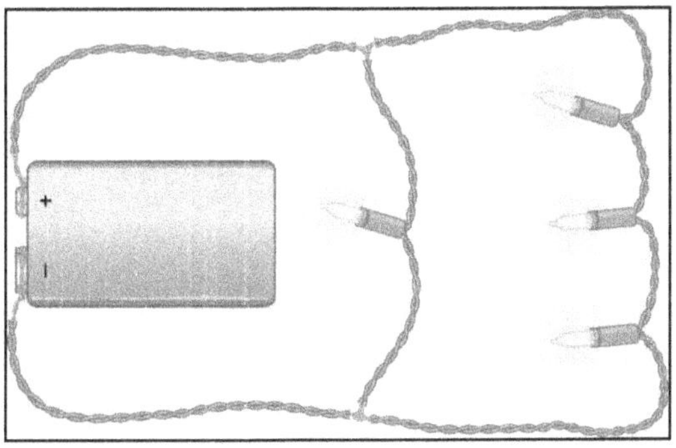

Figure 18.5: *This is one example of a complex circuit, but you may come up with your own design.*

Step 2: Use the meter to measure the voltage drop in each series and parallel part of your circuit. Enter the voltages for each part of your circuit in Table 18.3.

Step 3: Use the meter to measure the current in each series and parallel part of your circuit. Enter the currents for each part of your circuit in Table 18.3. *Caution—Do not connect both meter leads directly to the battery, since this can damage the meter.* Check with your teacher for specific instructions on how to set up your meter to get the correct current reading.

Data and Observations

Table 18.3 Voltage and Current in the Complex Circuit

	Voltage Drop (V)	Current (A)
Series part(s)		
Parallel part(s)		
Total Circuit		

Assessments

1. Determine the total voltage in your circuit. Explain your reasoning. TEKS (6)(C)
2. Determine the total current in your circuit. Explain your reasoning. TEKS (6)(C)
3. Calculate the power of your complex circuit. TEKS (6)(C)
4. Compare power consumption in your complex circuit with power consumption in your series and parallel circuits. Explain your answer. TEKS (6)(C)

Lab 19: Magnetism

Magnets have many uses. They create a tight seal in doors; they attract and pick up objects made of the metals iron, nickel, and cobalt; and in compasses they indicate which direction is north. One reason magnets are useful is that they exert **magnetic force** over a distance. The region over which a magnet exerts magnetic force is called a **magnetic force field**.

A magnet has two different ends—a **north pole** (N) and a **south pole** (S). A single pole cannot exist by itself. If a magnet is cut in half, the two halves each become a new magnet, with a north pole and a south pole. Depending on the alignment of the poles, magnets can attract or repel each other (Figure 19.1).

Figure 19.1: *A student investigates how two bar magnets interact. The two poles of each bar magnet are painted different colors.*

Although people use magnets often, magnetic substances are very rare. Of the millions of substances on Earth, only a few of them can be made into magnets. Why is this so? The reason has to do with the electrons in the atoms of magnetic substances. The movement of electrons in an atom causes each electron to act like a microscopic magnet. In most substances, the electrons' magnetic fields are not lined up in one direction, so there is no measurable magnetism. In substances that *are* magnetic, most of the magnetic fields are lined up in the same direction. This alignment makes these substances magnetic.

A magnet made from a magnetic substance is called a **permanent magnet**. Another kind of magnet, called an **electromagnet**, is created by moving electric **charges**, such as current in a wire. In this lab you will investigate both kinds of magnet.

<u>In this lab you will learn:</u>

- the direction of the magnetic force between two permanent magnets;
- factors that affect the strength of magnetic force;
- how to depict a magnetic force field;
- how current creates a magnetic force field.

Activity 1: Interactions of Magnets

When the poles of two magnets are brought near each other, they interact. In this activity, you measure both the direction and relative strength of the interactions between the poles of two permanent magnets.

Safety Precautions

- Wear proper eye protection.

For this activity you will need the following:

- A variety of bar magnets with poles marked
- A ruler marked in centimeters
- Masking tape
- Paper clips, all the same size (metal, not plastic)

For this activity you will work *in pairs*.

Procedure

Step 1: Place a pair of magnets near each other in two configurations: end to end and end to side. In which configuration is the magnetic force between the two magnets the strongest? In which configuration is the magnetic force the weakest? Record your findings in Table 19.1.

Step 2: Bring two north poles near each other. Is the force attractive or repulsive? Bring two south poles near each other. Is the force attractive or repulsive? Record your findings in Table 19.1.

Step 3: Bring a north pole and a south pole near each other. Is the force attractive or repulsive? Record your finding in Table 19.1.

Step 4: Position two magnets at the two ends of the ruler with their north poles facing each other. Slowly push one magnet toward the other as shown in Figure 19.2. Stop the magnet when you first feel a force.

a. At what distance do you first feel the repulsive force? Record the distance in Table 19.1.

b. Tape the magnet that is not being moved so it will not shift its position. Continue pushing the magnet. What happens to the strength of the repulsive force? Record your finding in Table 19.1.

Figure 19.2: *One magnet is being moved toward another magnet along a ruler.*

Step 5: Repeat Step 4, but flip one magnet so that its south pole faces the north pole of the other magnet. Slowly push one magnet toward the other. Stop the magnet when you first feel a force.

a. At what distance do you first feel the attractive force? Record the distance in Table 19.1.

b. Tape the magnet that is not being moved so it will not shift its position. Continue pushing the magnet. What happens to the strength of the attractive force? Record your finding in Table 19.1.

Step 6: Count how many paper clips each of the magnets can pick up. Is there a difference between the different magnets? Record your findings in Table 19.1.

Data and Observations

Table 19.1: Observations of Magnets

Step	Observation
1	
2	
3	
4	
5	
6	

Assessments

1. What is the general rule for the direction of magnetic force between magnetic poles? Refer to repulsion and attraction in your answer. TEKS (5)(G)
2. The interior of Earth acts like a large magnet. A compass needle is a magnet and the north pole of the needle points toward the geographic North Pole. Based on this, which one of Earth's magnetic poles is located near the geographic North Pole? Support your claim with evidence. TEKS (2)(G)
3. Review Figure 19.1. Why does the top magnet *float*? TEKS (2)(G)

Activity 2: Visualizing the Magnetic Force Field

You just saw that magnetic force is exerted over a distance. The area in which the force is exerted is called a *force field*. To show a force field, scientists use **lines of force**. The lines are not real, but they are useful for showing the direction and strength of the force field. The direction of force on the north pole of a magnet is shown by arrows on the lines. The relative strength of the magnet is shown by how close together the lines are. See the lines of force in Figure 19.3. In this activity, you visualize a magnetic force field.

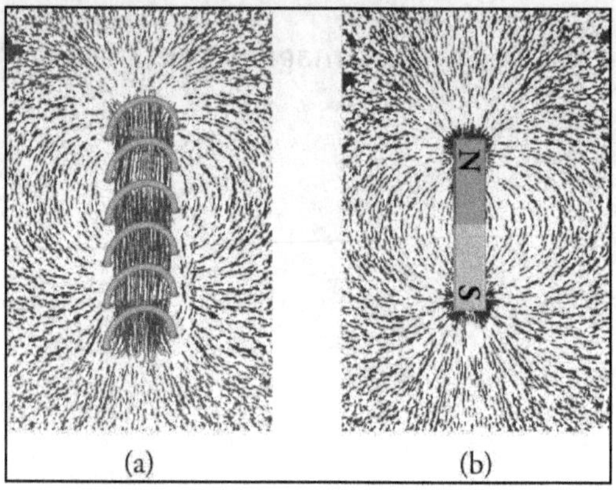

Figure 19.3: *Magnetic lines of force show the strength and direction of the magnetic field.*

Safety Precautions

- Wear proper eye protection.
- Wear a dust mask and gloves when handling iron filings. Handle the filings gently and carefully.
- Wash your hands thoroughly with soap and water before removing dust mask and eye protection.

For this activity you will need the following:

- One small bar magnet
- Iron filings stored in a closed container
- One piece of cardstock paper
- One small funnel

For this activity you will work *in groups of four*.

Procedure

Step 1: Fold the cardstock in half and press along the crease, then unfold the paper.

Step 2: Place the bar magnet on the table and place the cardstock on top of the magnet.

Step 3: Scatter the iron filings from the container carefully and evenly over the cardstock. Gently tap the paper in a variety of spots to separate the filings. *Note*—Do not allow the filings to touch the magnet. The filings are incredibly difficult to remove once attracted to a magnet.

Step 4: Study the pattern formed by the iron filings. Draw the pattern in your notebook.

Step 5: Using the crease in the stock card paper and the funnel, pour the iron filings carefully back into the container.

Assessments

1. You observed the magnetic force field on the cardstock, which is in the horizontal dimension. Does the force field also extend in all directions around the magnet? Support your claim with evidence. TEKS (5)(G)

2. The lines of force are closer together at the magnetic poles than they are in the center of the magnet. What does this indicate? TEKS (5)(G)

Activity 3: Building an Electromagnet

In their study of magnetism, scientists have learned that any moving charge generates a magnetic force field around it. An electromagnet is a magnet created by electric current moving through a wire. In this activity, you verify that an electromagnet acts the same as a permanent magnet; it has a north pole and a south pole and it exerts magnetic force. A simple electromagnet is shown in Figure 19.4.

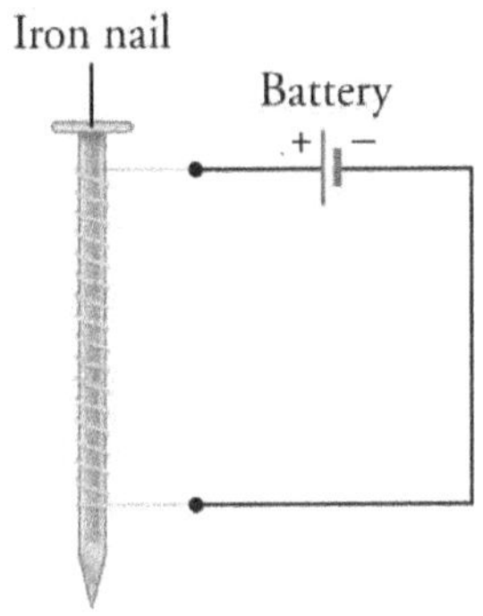

Figure 19.4: *You can produce an electromagnet by wrapping circuit wire around an iron nail and connecting the wire to a battery.*

<u>Safety Precautions</u>

- Wear proper eye protection.
- Be careful when handling sharp objects.
- Disconnect a clip lead immediately if the circuit smokes or gets hot.

<u>For this activity you will need the following</u>:

- One meter of fine-gauge circuit wire (copper or aluminum wire covered with thin insulation)
- One wire stripper
- Two D batteries and battery holder
- Two clip leads
- One magnet with poles labeled
- Twenty large metal paper clips (not plastic)
- Three galvanized iron nails
- Tape

For this activity you will work *in pairs*.

Procedure

Step 1: Use a wire stripper to remove the insulation from each end of the wire.

Step 2: Wrap the wire around one galvanized iron nail, placing the loops of the wrapped wire close together. Leave about 20 centimeters of straight wire on both ends.

Step 3: Use clip leads to attach the two ends of the wire to one D battery. This device is an electromagnet.

Step 4: Bring the bar magnet close to the electromagnet to determine the location of the electromagnet's north and south poles.

Step 5: Change the orientation of the battery so the direction of current moving through the electromagnet is in the opposite direction as it was in Step 4. Again bring the magnet close to the electromagnet to determine the location of the electromagnet's north pole and south pole. Do their locations change? Record your finding in Table 19.2.

Step 6: See how many paper clips the electromagnet can pick up. Record your finding in Table 19.2.

Step 7: Disconnect one of the clip leads to break the circuit. Add a second battery in series to the circuit and reconnect the circuit. See how many paper clips the electromagnet can pick up. Record your finding in Table 19.2.

Step 8: Disconnect one of the clip leads to break the circuit. Remove the second battery. Add the other two galvanized nails carefully to the center of the electromagnet. Loosen the loops of wire if necessary. Tape the nails together.

Step 9: Reconnect the circuit. Now test how many paper clips the electromagnet can pick up. Does the number change? Record your finding in Table 19.2.

Data and Observations

Table 19.2: Observations of an Electromagnet

Step	Observation
5	
6	
7	
8	

Assessments

1. You need to increase the magnetic strength of an electromagnet.

    a. What change would you make to the number of batteries?

    b. How does this change affect the current in the electromagnet?

    TEKS (5)(G)

2. A student investigating an electromagnet adds a second layer of wire loops around the nail in the same direction as the first loop. The electromagnet with more loops exerts a stronger magnetic force. Propose an explanation for this result. (*Hint*—Consider the number of electrons in the coil.) TEKS (5)(G)

3. In Assessment 2, what would happen to the strength of the magnetic force if the new loops were wrapped in the opposite direction as the first loop? Explain your answer.

Lab 20: The Atom

How are physicists able to determine which elements are present in the Sun? They obviously haven't traveled to the Sun to collect a sample. What they can do, however, is determine the **emission spectrum** of the Sun.

Every element has a unique **atomic spectrum** that allows scientists to determine which elements are present in a glowing object. The atomic spectrum is a pattern of electromagnetic radiation released or absorbed by atoms that have been excited energetically, such as by heat or an electronic voltage. Each atomic spectrum is unique to a particular element, showing a pattern of radiation determined by its atomic structure. If the atom is radiating energy, it produces an emission spectrum of bright lines. If it is absorbing energy, it produces the same pattern, but consists of dark lines superimposed on the spectrum of the light source (called an *absorption spectrum).* Knowing what the atomic spectra of hydrogen and helium are and then looking at the emission spectrum of the Sun allowed scientists to determine that those elements are present within the Sun. If there is a mixture of gases (as there is in the Sun), scientists can determine which atomic spectra are being combined to form the observed emission spectrum.

The reason emission spectra are produced has to do with **electrons** *jumping down* to a lower **energy level**. When an electron loses energy by moving to a lower energy level, it releases that energy in the form of a **photon**. The photon leaves the atom with the same amount of energy that the electron lost. Therefore, the larger the jump an electron makes, the greater the energy of the photon. Because photon energy is related directly to the frequency of the photon, and the frequency of the photon can tell us the wavelength, or color, of the photon, each jump produces a very specific color of light. No two elements produce the same atomic spectra, which makes the atomic spectrum of any given element like a *fingerprint* for that element.

<u>In this lab you will learn:</u>

- how to compare the atomic spectra produced by different elements;
- how to explain how emission spectra are produced by electrons in the atom.

Activity 1: Emission Spectrum TEKS (8)(B)

Detectives can use fingerprints to determine which person may have committed a crime because fingerprints are unique to every person. This is similar to how a scientist can use **emission spectra** to determine which **elements** or atoms are present in a glowing object. Every atom has a unique **atomic spectrum** that identifies its presence within an object. When multiple elements are present in a gas, the emission spectrum of that gas is a combination of the atomic spectra produced by all the elements the gas contains. For this activity you use a digital spectroscope to observe and draw emission spectra for various pure gases present in a **spectrum tube**. (This may be a familiar procedure if you did the experiment on light and color.) In this activity you identify the elements present in mixed gases by comparing their emission spectra with the emission spectrum you drew.

Safety Precautions

- Handle the spectroscope with care; it is fragile.
- Don't stare at the spectrum tubes for too long; this is similar to staring at a light bulb.
- The spectrum tubes can become very hot, so turn them off and let them cool before handling them. Use a thermal glove to handle any material that may be hot.

For this activity you will need the following:

- A digital spectroscope (either one per group or one for the entire class)
- Spectrum tubes of various elements
- Graph paper and colored pencils

For this activity you will work *in pairs* if there are enough spectroscopes; otherwise, if there are fewer spectroscopes, you will work as a *group or as a class*. Listen closely to your teacher's instructions.

Procedure

Step 1: Follow your teacher's instructions carefully in setting up the digital spectroscope carefully and turning it on (Figure 20.1). Ask your teacher for help if you are unsure how to perform this task. Point the spectroscope toward the spectrum tube, but don't turn on the spectrum tube just yet.

Figure 20.1: *This is one type of digital spectroscope that you may be using. Be sure to avoid touching the lens portion of the instrument.*

Step 2: Turn on the spectrum tube of the first element and make sure your digital spectroscope is pointed directly at it. You should see specific colors appear on your display. In the following Data and Observations section, use the colored pencils to draw the specific colored bars at their respective wavelengths. The colored pencils might not match the colors on the screen perfectly, but pick a color that is close.

Step 3: Turn off the spectrum tube and allow it time to cool before handling it. As a precaution, use a thermal glove to handle materials that may be hot. When the spectrum tube is cool enough to handle, obtain a different element and repeat Step 2 with the new element. Repeat this step for all the spectrum tubes you have in the class.

Step 4: Your teacher will now present you with a tube that has a mixture of gases. Your job is to use the emission spectra you drew in Step 3 and Step 4 to identify which elements are present in this unknown gas tube. Complete this task by drawing the emission spectrum of the mixed gas tube on your graph paper and then analyze it using the atomic spectrum for each element that you drew previously and recorded in the following section.

Data and Observations

Name of Element:

400 450 500 550 600 650 700

Wavelength (nm)

Name of Element:

400 450 500 550 600 650 700 400 450

Wavelength (nm)

Name of Element:

400 450 500 550 600 650 700 400 450

Wavelength (nm)

Name of Element:

400 450 500 550 600 650 700 400 450

Wavelength (nm)

Mixed-Gas Tube

400 450 500 550 600 650 700 400 450

Wavelength (nm)

Which elements are present in the mixed-gas tube? Explain why you think this is the case. (Write your answers below.)

Assessments

1. Can you use an emission spectrum to determine which elements are present in a cold object? Why or why not? TEKS (8)(B)
2. Why did the wavelength on the graph paper on which you drew your emission spectrum range from 400 nanometers to 700 nanometers? Why not higher or lower than these two values? TEKS (8)(B)
3. Design a procedure to determine the composition of the gas within a fluorescent lighting tube. TEKS (8)(B)

Activity 2: Electron Energy Transitions

Why does each element have its own unique atomic spectrum? What is happening inside the atom that causes this effect? The color bands are actually a result of photons emitted at very specific wavelengths. Each wavelength refers to a specific color in the visible portion of the electromagnetic spectrum. The photons have this wavelength because of the amount of energy they possess. Where does the photon get this energy? The energy is released by electrons moving from a high energy level around the nucleus to a lower energy level (Figure 20.2). Similar to a ball turning potential energy into kinetic energy as it rolls down the hill, as an electron *jumps down* to a lower energy level, it turns that energy into photons. In this activity you use emission spectra to determine the number of electron **energy transitions** made in an atom.

Safety Precautions

- Handle the spectroscope with care; it is fragile.
- Do not stare at the spectrum tubes for too long; this is similar to staring at a light bulb.
- The spectrum tubes can become very hot, so turn them off and let them cool before handling them. Use a thermal glove to handle any material that may be hot.

For this activity you will need the following:

- A digital spectroscope (either one per group or one for the entire class)
- A hydrogen spectrum tube
- Colored pencils

For this activity you will work *in pairs* if there are enough spectroscopes; otherwise, if there are fewer spectroscopes, you will work as a group or *as a class*. Listen closely to your teacher's instructions.

Procedure

Step 1: Follow your teacher's instructions carefully in setting up the digital spectroscope carefully and turning it on. Ask your teacher for help if you are unsure how to do these tasks. Point the spectroscope toward the spectrum tube, but don't turn the spectrum tube on just yet.

Step 2: Turn on the spectrum tube and make sure your digital spectroscope is pointed directly at it. You should see specific colors appear on your display. In the following Data and Observations section, use the colored pencils and draw the specific colored bars at their respective wavelengths. The colored pencils might not match the colors on the screen perfectly, but pick a color that is close. Also, list the values of the wavelengths in Table 20.1.

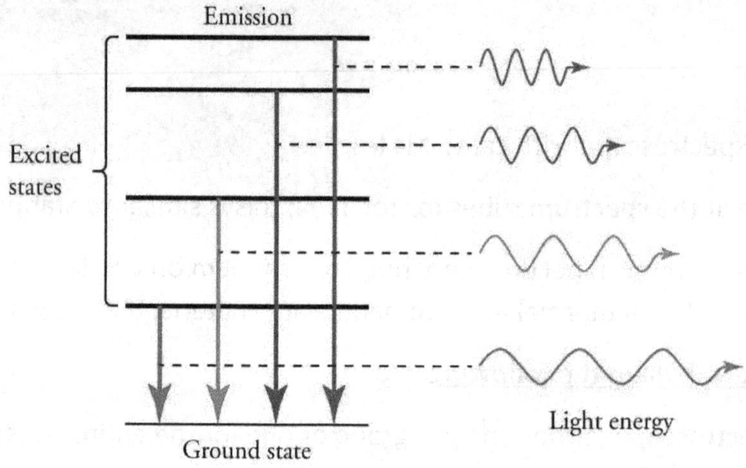

Figure 20.2: *As an electron moves down a level, it emits a photon of a specific wavelength.*

Step 3: Using the equation $E = hc / \lambda$, convert the wavelengths you recorded into the energy the photon possesses. Record these values in the appropriate spots in Table 20.1.

Data and Observations

Hydrogen Spectrum Tube

400 450 500 550 600 650 700 400 450

Wavelength (nm)

Table 20.1: Photon Wavelengths and Energies

Band Number	Color of Band	Wavelength of Photon	Energy of Photon
1			
2			
3			
4			

Assessments

1. If each element has a specific emission spectrum, what does this tell you about the electron energy levels among atoms? TEKS (8)(B)
2. How is wavelength related to the energy change of an electron when it moves from an excited state to a lower state? TEKS (8)(B)
3. Pick two of the atomic spectra that you drew in Activity 1. How do the electron energy levels differ between the hydrogen atom and the elements you picked? Justify your answer using evidence from the spectra. TEKS (8)(B)